S0-BBR-257

Mathematics for Liberal Arts Majors

Mathematics for Liberal Arts Majors

Christopher Thomas

Schaum's Outline Series

New York   Chicago   San Francisco   Lisbon   London   Madrid   Mexico City
Milan   New Delhi   San Juan   Seoul   Singapore   Sydney   Toronto

The **McGraw·Hill** Companies

CHRISTOPHER THOMAS is professor of math at the Massachusetts College of Liberal Arts. He is the author of *Calculus Success in 20 Minutes a Day* (2006) and *Trigonometry Success in 20 Minutes a Day* (2007). Christopher has been teaching the Math for Liberal Arts course for the past four years.

Schaum's Outline of
MATHEMATICS FOR LIBERAL ARTS MAJORS

Copyright © 2009 by The McGraw-Hill Companies, Inc. All rights reserved. Printed in the United States of America. Except as permitted under the United States Copyright Act of 1976, no part of this publication may be reproduced or distributed in any form or by any means, or stored in a database or retrieval system, without the prior written permission of the Publisher.

1 2 3 4 5 6 7 8 9 CUS/CUS 0 1 4 3 2 1 0 8

ISBN: 978-0-07-154429-0
MHID: 0-07-154429-1

McGraw-Hill books are available at special quantity discounts for use as premiums and sales promotions, or for use in corporate training programs. For more information, please write to the Director of Special Sales, McGraw-Hill Professional, Two Penn Plaza, New York, NY, 10121-2298. Or contact your local bookstore.

Sponsoring Editor: Anya Kozorez
Production Editor: Tama L. Harris
Editing Supervisor: Frank Kotowski, Jr.

Library of Congress Cataloging-in-Publication Data

Thomas, Christopher, 1973–
    Schaum's outline of mathematics for liberal arts majors / Christopher Thomas.—1st ed.
      p. cm.—(Schaum's outline series)
    Includes index.
    ISBN-13: 978-0-07-154429-0
    ISBN-10: 0-07-154429-1
    1. Mathematics–Outlines, syllabi, etc. I. Title. II. Title: Mathematics for liberal arts majors.

QA37.3.T46 2008
510–dc22
                                                      2008018629

R042984801३

Contents

Mathematics for
Liberal Arts Majors

CHAPTER 1

Number Systems

In this chapter we will study different *number systems*, ways of writing numbers, from those of ancient history to those of the modern day.

The Base-Ten Decimal System

To begin comparing number systems, let us start with a review of the base-ten decimal system. Base-ten numbers are written using ten symbols which, like fingers, are called *digits*: 0, 1, 2, 3, 4, 5, 6, 7, 8, and 9. Base-ten is a *positional* system in that the location relative to the *decimal point* indicates the power of ten represented by each digit, as illustrated in Fig. 1-1.

$$10^4 \qquad 10^3 \qquad 10^2 \qquad 10^1 \qquad 10^0 \qquad 10^{-1} \qquad 10^{-2} \qquad 10^{-3}$$

$$10{,}000\text{'s} \quad 1{,}000\text{'s} \quad 100\text{'s} \quad 10\text{'s} \quad 1\text{'s} \quad \tfrac{1}{10}\text{'s} \quad \tfrac{1}{100}\text{'s} \quad \tfrac{1}{1.000}\text{'s}$$

Fig. 1-1

As an example, the number 385.16 represents $300 + 80 + 5 + \dfrac{1}{10} + \dfrac{6}{100}$. When it is necessary to specify that a number is written in base-ten, a subscript 10 is used, for example 385.16_{10}.

The base-ten digits have long been known as *Arabic numerals* because the Europeans copied them from the Muslims of Spain during the Crusades. When it was discovered that the Arabs had copied the system from the people of India, the name was updated to *Hindu-Arabic numerals*. Recent speculation has further suggested that the Indians may have been influenced by the ancient Chinese, who also used a base-ten positional system.

Tally Marks

One of the oldest artifacts of human mathematics is a wolf bone from 30,000 b.c. with many straight notches cut into it. This is an example of *tally marks*, the very first numbering system. The number 1 is represented by a vertical line, and every successive number is represented by an additional vertical line. The first seven numbers are written in tally marks in Fig. 1-2.

Fig. 1-2

This system was improved by separating the marks into groups of five. This was done by putting every fifth mark diagonally across the last four. In Fig. 1-3, the first seven numbers are written in this form.

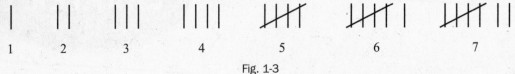

Fig. 1-3

Tally marks are still used today. Perhaps not coincidentally, the modern number 1 is still represented by a single vertical line. An advantage unique to tally marks is that increasing a number never requires erasing. Thus, they are very useful for keeping track of small numbers that periodically increase, like scores or votes. However, it would be ridiculous to try to read or record large numbers with tally marks.

SOLVED PROBLEMS

Tally Marks

1.   Write the numbers (a) 8, (b) 12, (c) 15, and (d) 23 with both kinds of tally marks.
2.   Convert the numbers in Fig. 1-4 into base-ten.

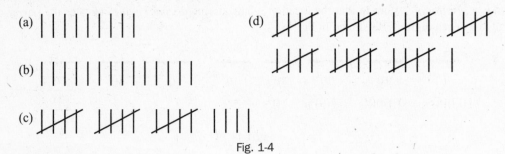

Fig. 1-4

Answers

1.   Shown in Fig. 1-5.
2.   (a) 9, (b) 14, (c) 19, and (d) 36

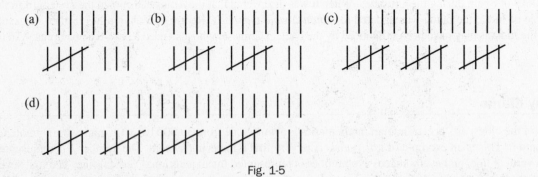

Fig. 1-5

Egyptian Numbers

A next step toward a better number system was developed by the ancient Egyptians. They made hieroglyphs to represent each of the powers of ten, as illustrated in Fig. 1-6. This way no more than 9 of any symbol would ever be needed in a number.

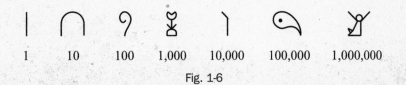

Fig. 1-6

For example, the number 5,286 would be represented as in Fig. 1-7. The order of the symbols does not matter, but traditionally the smaller numbers were put first.

Fig. 1-7

To represent unit fractions, the ancient Egyptians would put an oval above the number representing the denominator. For examples, $\frac{1}{4}, \frac{1}{25}, \frac{1}{100}$, and $\frac{1}{1,312}$ are all illustrated in Fig. 1-8.

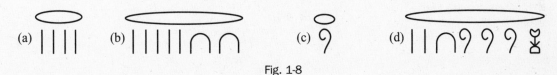

Fig. 1-8

For some reason, the Egyptians would only use unit fractions and never use the same denominator twice. For example, the quantity $\frac{2}{5}$ would not be represented by $\frac{1}{5}+\frac{1}{5}$, but by $\frac{1}{3}+\frac{1}{15}$ or $\frac{1}{4}+\frac{1}{10}+\frac{1}{20}$. This made working with fractions very difficult.

SOLVED PROBLEMS

Egyptian Numbers

1. Use the ancient Egyptian numbering system to write (a) 8, (b) 35, (c) 460, (d) 1,225, (e) 6,341,281, (f) $\frac{1}{3}$, (g) $\frac{1}{125}$, and (h) $\frac{1}{1,000}$.

2. Translate the numbers in Fig. 1-9 into base-ten.

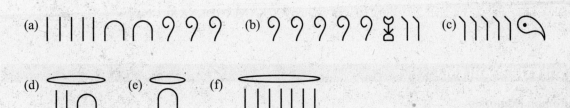

Fig. 1-9

Answers

1. Shown in Fig. 1-10.
2. (a) 325, (b) 21,500, (c) 150,000, (d) $\frac{1}{12}$, (e) $\frac{1}{10}$, and (f) $\frac{1}{7}$

Fig. 1-10

Roman Numerals

The next step after the Egyptian number system (in concept, not in time) is the system of *Roman numerals*. Just as with the Egyptians, there is a different symbol for the first few powers of ten: $1=I$, $10=X$, $100=C$, and $1,000=M$. In addition, there are symbols for $5=V$, $50=L$, and $500=D$.

At first, the Roman number system was essentially the same as the Egyptian one except that there were more symbols and the largest ones appeared on the left. For example, at one time the number 49 was written XXXXVIIII and the number 2,864 was MMDCCCLXIIII. Later on a shortcut was introduced:

> If *I* comes before *V* or *X*, then it subtracts 1 from the number.
> If *X* comes before *L* or *C*, then it subtracts 10 from the number.
> If *C* comes before *D* or *M*, then it subtracts 100 from the number.

For example, instead of XXXXVIIII, the number 49 would be written XLIX, meaning $10+50+(-1)+10$. Similarly, 1995 would be written MCMXCV, meaning $1,000-100+1,000-10+100+5$.

To represent larger numbers, bars could be placed over some of the numerals. Each bar represented multiplication by 1,000. For example, the number 5,000 was written $\overline{V}$, the number 84,000,000 was written $\overline{\overline{LXXXIV}}$, and the number 239,480,216 was written $\overline{\overline{CCXXXIX}}\overline{CDLXXX}CCXVI$.

Roman Numerals

1. Convert the following numbers from Roman numerals into modern numbers:

   (a) XXXIII
   (b) IV
   (c) CCXCV
   (d) MMMCDLVIII
   (e) CMXCIX

(f) $\overline{V}CCLXXX$

(g) $\overline{IV}XCVI$

(h) $\overline{\overline{DCCCXXIX}}CCXLDCCLII$

2. Write the following numbers using the Roman number system:

(a) 300
(b) 54
(c) 2,007
(d) 479
(e) 1,982
(f) 15,291
(g) 10,480
(h) 89,360,219

Answers

1. (a) 33, (b) 4, (c) 295, (d) 3,458, (e) 999, (f) 5,280, (g) 4,096, and (h) 829,240,752

2. (a) CCC, (b) LIV, (c) MMVII, (d) CDLXXIX, (e) MCMLXXXII, (f) $\overline{XV}CCXCI$, (g) $\overline{X}CDLXXX$, and

(h) $\overline{\overline{LXXXIX}}CCCLXCCXIX$

The Babylonian Number System

The number system used in Babylon and the other ancient cities of Mesopotamia was more sophisticated than the Roman system even though it predated the Romans by thousands of years.

The Babylonian system used only two symbols: a vertical line with a wedge on top representing the number 1 and a triangle representing the number 10, as illustrated in Fig. 1-11(a). For numbers up to 59, these symbols were used just as they would be under the Egyptian system. For example, the number 17 is illustrated in Fig. 1-11(b), the number 23 is in Fig. 1-11(c), and the number 45 is in Fig. 1-11(d).

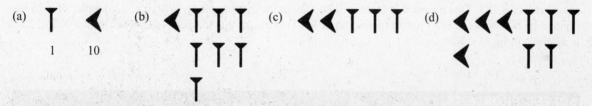

Fig. 1-11

For numbers beyond 59, the Babylonians used a base-sixty positional system. Rather than each place representing a power of ten, as in the base-ten system, each place represented a power of 60, as illustrated in Fig. 1-12.

60^3	60^2	60^1	60^0	60^{-1}	60^{-2}
216,000's	3600's	60's	1's	$\frac{1}{60}$'s	$\frac{1}{3,600}$'s

Fig. 1-12

There was no zero in the Babylonian system, so they would leave a gap to indicate when a place was empty. As well, there was no decimal point in the Babylonian system, so there was no way to be sure what each place represented. It was assumed that a reader would have some idea as to the scale of the thing being measured, thus whether the first place represented 1's, 60's, 3,600's, or more. For our purposes here, we will put a marker to the right of the 1's place to avoid confusion.

For example, the number in Fig. 1-13(a) has 25 in the 60's place and 32 in the 1's place, and thus represents the number $25 \times 60 + 32 \times 1 = 1,532$. The number in Fig. 1-13(b) has 1 in the 1's place, 24 in the $\frac{1}{60}$'s place, 51 in the $\frac{1}{3,600}$'s place, and 10 in the $\frac{1}{60^3} = \frac{1}{216,000}$'s place. This number, $1 + \frac{24}{60} + \frac{51}{3,600} + \frac{10}{216,000} \approx 1.414213$, was the Babylonian approximation of $\sqrt{2}$, accurate to six decimal places.

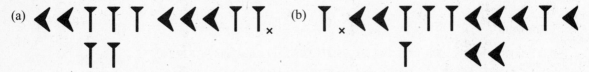

Fig. 1-13

To convert a number into base-sixty, evaluate the value of the highest possible place first. For example, the number $1,000_{10}$ is bigger than 60 but less than 3,600. There will thus be no numbers in the 3,600's place. When 1,000 is divided by 60, the result is 16 with a remainder of 40. (Hint: on a calculator, $1,000 \div 60 \approx 16.667$; thus, the answer is 16. The remainder is found by calculating $1,000 - 16 \times 60 = 40$.) Thus, the Babylonian number for $1,000_{10}$ has 16 in the 60's place and 40 in the 1's place, as illustrated in Fig. 1-14(a).

Fig. 1-14

The number $75,283_{10}$ is bigger than 3,600 but less than 216,000, so we divide by 3,600 and get 20 with a remainder of 3,283. This means that there will be a 20 in the 3,600's place. When the remainder of 3,283 is divided by 60, the result is 54 with a remainder of 43. This means that the Babylonian equivalent of $75,283_{10}$ is as given in Fig. 1-14(b).

SOLVED PROBLEMS

Babylonian Numbers

1. Convert each of the numbers in Fig. 1-15 into base-ten.
2. Use the Babylonian number system to write the numbers (a) 34, (b) 256, (c) 5,300, and (d) 1,000,000.

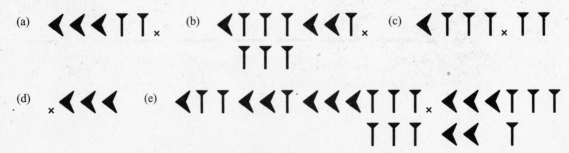

Fig. 1-15

Answers

1. (a) 32

   (b) $16 \times 60 + 21 = 981$

   (c) $13 + \dfrac{2}{60} = 13.03333... = 13.0\overline{3}$

   (d) $\dfrac{30}{60} = 0.5$

   (e) $12 \times 3,600 + 21 \times 60 + 36 + \dfrac{54}{60} = 44,496.9$

2. (a) Shown in Fig. 1-16(a).

   (b) On a calculator, $256 \div 60 = 4.2666... = 4.2\overline{6}$, which means 4 with a remainder of $256 - 4 \times 60 = 16$. Thus, this number has a 4 in the 60's place and a 16 in the 1's place, as shown in Fig. 1-16(b).

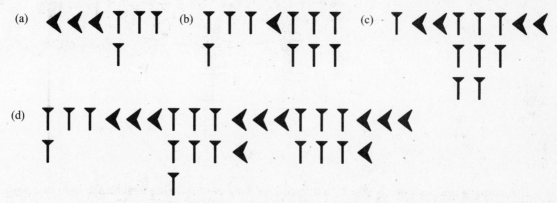

Fig. 1-16

   (c) 5,300 has one 3,600 with a remainder of 1,700. This remainder, when divided by 60, results in 28 with a remainder of 20. Thus this number, in the Babylonian number system, has a 1 in the 3,600's place, a 28 in the 60's place, and a 20 in the 1's place, as shown in Fig. 1-16(c).

   (d) The largest power of 60 in 1,000,000 is 216,000, which goes in 4 times with a remainder of $1,000,000 - 4 \times 216,000 = 136,000$. When 136,000 is divided by 3,600, the result is 37 with a remainder of $136,000 - 37 \times 3,600 = 2,800$. When 2,800 is divided by 60, the result is 46 with a remainder of $2,800 - 46 \times 60 = 40$. Thus, $1,000,000_{10}$ will have a 4 in the 216,000's place, a 37 in the 3,600's place, a 46 in the 60's place, and a 40 in the 1's place, as shown in Fig. 1-16(d).

Binary Numbers

The *binary* or *base-two* number system is a positional system where each place represents a power of 2, as illustrated in Fig. 1-17. There are only two digits: 0 and 1. Sometimes a subscript of 2 is used to indicate that a number is written in base-two. Spaces are sometimes placed after every fourth place to make long binary numbers easier to read.

2^7	2^6	2^5	2^4	2^3	2^2	2^1	2^0	2^{-1}	2^{-2}	2^{-3}
128's	64's	32's	16's	8's	4's	2's	1's	$\dfrac{1}{2}$'s	$\dfrac{1}{4}$'s	$\dfrac{1}{8}$'s

Fig. 1-17

For example, $10\ 1101_2$ has a 1 in the 32's place, a 1 in the 8's place, a 1 in the 4's place, and a 1 in the 1's place. Thus $10\ 1101_2 = 32 + 8 + 4 + 1 = 45_{10}$. Similarly, the number 1001.101_2 has a 1 in the 8's place, a 1 in the 1's place, a 1 in the $\frac{1}{2}$'s place, and a 1 in the $\frac{1}{8}$'s place. Thus, $1001.101_2 = 8 + 1 + \frac{1}{2} + \frac{1}{8} = 9.625_{10}$.

To convert a number from base-ten to base-two, repeatedly subtract the largest power of 2 smaller than the current remainder. For example, the largest power of 2 less than 105_{10} is 64. When this is subtracted, the remainder is 41. The largest power of 2 in 41 is 32. When this is subtracted, the remainder is 9. The largest power of 2 in 9 is 8, with a remainder of 1. This means that $105_{10} = 64 + 32 + 8 + 1 = 110\ 1001_2$. When this computation is done on paper, the result looks like Fig. 1-18(a). Similarly, Fig. 1-18(b) shows the computations necessary for converting 598_{10} into base-two: $598_{10} = 512 + 64 + 16 + 4 + 2 = 10\ 0101\ 0110_2$. To make this process even easier, make a list of all the powers of 2 somewhere on the paper (1, 2, 4, 8, 16, 32, ...); keep doubling until you reach numbers larger than you will need.

$$
\begin{array}{rr}
\text{(a)} & 105 \\
& -64 \\
\hline
& 41 \\
& -32 \\
\hline
& 9 \\
& -8 \\
\hline
& 1
\end{array}
\qquad
\begin{array}{rr}
\text{(b)} & 598 \\
& -512 \\
\hline
& 86 \\
& -64 \\
\hline
& 22 \\
& -16 \\
\hline
& 6 \\
& -4 \\
\hline
& 2
\end{array}
$$

Fig. 1-18

The memory in a computer is kept by a tremendous number of switches which each have two positions: on (1) and off (0). Because of this, all computer operations are ultimately performed with binary numbers.

SOLVED PROBLEMS

Binary Numbers

1. Convert the following numbers into base-ten: (a) 11_2, (b) 1010_2, (c) 1111_2, (d) $1000\ 1101_2$, (e) $1110\ 1011\ 0111_2$, (f) $101.\ 11_2$, and (g) $10\ 0011.0010\ 1_2$.

2. Convert the following numbers into base-two: (a) 7_{10}, (b) 30_{10}, (c) 87_{10}, (d) 194_{10}, (e) 444_{10}, and (f) $1{,}000_{10}$.

Answers

1.   (a) $11_2 = 2 + 1 = 3_{10}$

     (b) $1010_2 = 8 + 2 = 10_{10}$

     (c) $1111_2 = 8 + 4 + 2 + 1 = 15_{10}$

     (d) $1000\ 110_2 = 128 + 8 + 4 + 1 = 141_{10}$

     (e) $1110\ 1011\ 0111_2 = 2{,}048 + 1{,}024 + 512 + 128 + 32 + 16 + 4 + 2 + 1 = 3{,}767_{10}$

     (f) $101.11_2 = 4 + 1 + \frac{1}{2} + \frac{1}{4} = 5.75_{10}$

     (g) $10\ 0011.0010\ 1_2 = 32 + 2 + 1 + \frac{1}{8} + \frac{1}{32} = 35.15625_{10}$

2.   The computations are shown in Fig. 1-19.

(a)	(b)	(c)	(d)	(e)	(f)
7	30	87	194	444	1,000
−4	−16	−64	−128	−256	−512
3	14	23	66	188	488
−2	−8	−16	−64	−128	−256
1	6	7	2	60	232
	−4	−4		−32	−128
	2	3		28	104
		−2		−16	−64
		1		12	40
				−8	−32
				4	8

Fig. 1-19

(a) $7_{10} = 4 + 2 + 1 = 111_2$
(b) $30_{10} = 16 + 8 + 4 + 2 = 1\ 1110_2$
(c) $87_{10} = 64 + 16 + 4 + 2 + 1 = 101\ 0111_2$
(d) $194_{10} = 128 + 64 + 2 = 1100\ 0010_2$
(e) $444_{10} = 256 + 128 + 32 + 16 + 8 + 4 = 1\ 1011\ 1100_2$
(f) $1,000_{10} = 512 + 256 + 128 + 64 + 32 + 8 = 11\ 1110\ 1000_2$

Hexadecimal Numbers

When programming computers, the long sequences of 0's and 1's necessary for binary numbers can become very tedious. To abbreviate matters, programmers often use a *hexadecimal* or *base-sixteen* number system. The 16 symbols used to represent numbers are 0, 1, 2, 3, 4, 5, 6, 7, 8, 9, A = 10_{10}, B = 11_{10}, C = 12_{10}, D = 13_{10}, E = 14_{10}, and F = 15_{10}. As usual, each place represents a different power of 16, as illustrated in Fig. 1-20. A subscript of 16 is used to indicate a hexadecimal number

16^4	16^3	16^2	16^1	16^0	16^{-1}	16^{-2}
65,536's	4,096's	256's	16's	1's	$\frac{1}{16}$'s	$\frac{1}{256}$'s

Fig. 1-20

For example, the base-sixteen number $B2_{16}$ represents $11 \times 16 + 2 \times 1 = 178_{10}$. Similarly, $F8C_{16} = 15 \times 256 + 8 \times 16 + 12 \times 1 = 3,980_{10}$, and $1A.7D_{16} = 1 \times 16 + 10 \times 1 + \frac{7}{16} + \frac{13}{256} \approx 26.488_{10}$.

The process of converting numbers from base-ten to hexadecimal is similar to that of the Babylonian base-sixty numbers: divide the number by the highest power of 16 possible and then repeat the process on the remainder. For example, the number $1,000_{10}$ is bigger than 256 but less than 4,096, so we divide by 256. When we divide 1,000 by 256, we get 3 with a remainder of $1,000 - 3 \times 256 = 232$. When 232 is divided by 16, the result is 14 with a remainder of $232 - 14 \times 16 = 8$. The symbol for 14 is E, so $1,000_{10} = 3E8_{16}$.

Switching between hexadecimal and binary numbers is very easy when the 16 hexadecimal digits are put into 4-digit binary blocks: $0_{16} = 0000_2$, $1_{16} = 0001_2$, $2_{16} = 0010_2$, $3_{16} = 0011_2$, $4_{16} = 0100_2$, $5_{16} = 0101_2$, $6_{16} = 0110_2$, $7_{16} = 0111_2$, $8_{16} = 1000_2$, $9_{16} = 1001_2$, $A_{16} = 1010_2$, $B_{16} = 1011_2$, $C_{16} = 1100_2$, $D_{16} = 1101_2$, $E_{16} = 1110_2$, and $F_{16} = 1111_2$. For example, the number $F8_{16}$ in binary is formed by putting together $F_{16} = 1111_2$ and $8_{16} = 1000_2$, thus $F8_{16} = 1111\ 1000_2$.

Converting from binary to hexadecimal is just as easy. For example, the number $110\ 1011\ 0011_2$ can be viewed as $6_{16} = 0110_2$, $B_{16} = 1011_2$, and $3_{16} = 0011_2$ put together, so $110\ 1011\ 0011_2 = 6B3_{16}$.

SOLVED PROBLEMS

Hexadecimal Numbers

1. Convert into base-ten numbers: (a) $2D_{16}$, (b) BC_{16}, (c) $310A_{16}$, and (d) $F9.82_{16}$.
2. Convert into hexadecimal numbers: (a) 100_{10}, (b) $5,280_{10}$, (c) $1110\ 1110_2$, and (d) $1001\ 0110\ 1010_2$.
3. Convert into binary: (a) $2F_{16}$, (b) $70B_{16}$, (c) 416_{16}, and (d) $COFFEE_{16}$.

Answers

1. (a) $2D_{16} = 2 \times 16 + 13 \times 1 = 45_{10}$
   (b) $BC_{16} = 11 \times 16 + 12 \times 1 = 188_{10}$
   (c) $310A_{16} = 3 \times 4,096 + 1 \times 256 + 10 \times 1 = 12,554_{10}$
   (d) $F9.82_{16} = 15 \times 16 + 9 \times 1 + \dfrac{8}{16} + \dfrac{2}{256} = 249.5078125_{10}$

2. (a) 100_{10} divided by 16 is 6, remainder $100 - 6 \times 16 = 4$, thus $100_{10} = 64_{16}$.
   (b) $5,280_{10}$ divided by 4,096 is 1, remainder $5,280 - 4,096 = 1,184$. Thus 1,184, divided by 256, is 4, remainder $1,184 - 4 \times 256 = 160$. Thus $160 = 10 \times 16$. Because A is the symbol for 10, this means $5,280_{10} = 14A0_{16}$.
   (c) $1110\ 1110_2$, broken into 4-digit blocks, is 1110 and 1110. Because $1110_2 = E_{16}$, this means that $1110\ 1110_2 = EE_{16}$.
   (d) $1001\ 0110\ 1010_2$ breaks into 1001, 0110, and 1010; thus, $1001\ 0110\ 1010_2 = 96A_{16}$ because $9_{16} = 1001_2$, $6_{16} = 0110_2$, and $A_{16} = 1010_2$.

3. (a) $2F_{16} = 0010\ 1111_2 = 10\ 1111_2$
   (b) $70B_{16} = 0111\ 0000\ 1011_2 = 111\ 0000\ 1011_2$
   (c) $416_{16} = 0100\ 0001\ 0110_2 = 100\ 0001\ 0110_2$
   (d) $COFFEE_{16} = 1100\ 0000\ 1111\ 1111\ 1110\ 1110_2$

A Fast Algorithm for Changing Bases

To convert a number into base b, divide it by b. The remainder will be the 1's digit of the number in base b. When the quotient is divided by b, the remainder will be the b's digit of the number in base b. When the new quotient is divided by b, the remainder will be the b^2's digit of the number in base b. Repeat this process until the quotient is zero.

For example, suppose we want to convert 75_{10} into base 2. When 75 is divided by 2, the quotient is 37 with a remainder of 1. When 37 is divided by 2, the quotient is 18 with a remainder of 1. If the calculations are made as illustrated in Fig. 1-21, the answer is found by reading down the list of remainders. Thus, $75_{10} = 100\ 1011_2$.

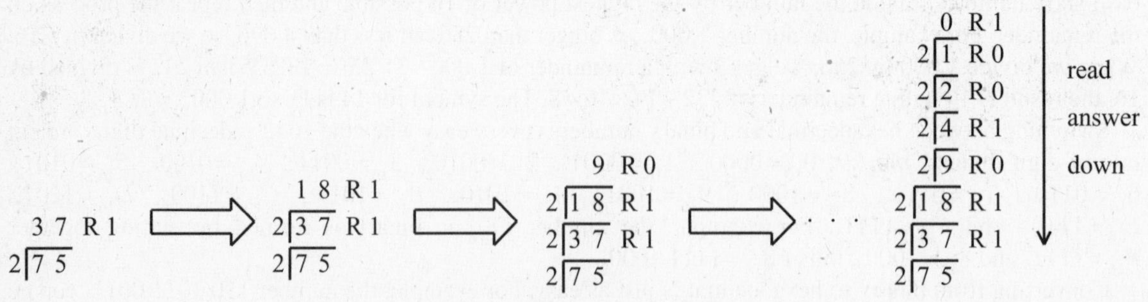

Fig. 1-21

SOLVED PROBLEMS

A Fast Algorithm for Changing Bases

1. Convert 85_{10} into base 2.
2. Convert 126_{10} into base 4.
3. Convert 3248_{10} into base 8.
4. Convert 3250_{10} into base 16.

Answers

1. It can be seen that $85_{10} = 1010101_2$ using the calculations in Fig. 1-22(a).
2. The calculations by which the answer 1332_4 is obtained are shown in Fig. 1-22(b).

```
      0   R 1
    2⌐1   R 0
    2⌐2   R 1
    2⌐5   R 0
   2⌐1 0  R 1         0   R 1          0   R 6
   2⌐2 1  R 0        4⌐1  R 3         8⌐6  R 2                  0   R 12
   2⌐4 2  R 1        4⌐7  R 3        8⌐5 0 R 6           16 ⌐1 2  R 11
   2⌐8 5              4⌐3 1 R 2      8⌐4 0 6 R 0         16 ⌐2 0 3 R 2
                     4⌐1 2 6         8⌐3 2 4 8           16 ⌐3 2 5 0

     (a)               (b)             (c)                 (d)
```

Fig. 1-22

3. As shown in Fig. 1-22(c), $3248_{10} = 6260_8$.
4. In Fig. 1-22(d), we see that 3250_{10} has only 3 digits in base 16: the 12th digit (C), the 11th digit (B), and the second digit (2). Thus, $3250_{10} = CB2_{16}$.

SUPPLEMENTAL PROBLEMS

1. Count to 21 in (a) base-ten, (b) tally marks, (c) Egyptian numbers, (d) Roman numerals, (e) Babylonian numbers, (f) binary, and (g) base-sixteen.
2. Write the number 42_{10} in (a) tally marks, (b) Egyptian numbers, (c) Roman numerals, (d) Babylonian numbers, (e) binary, and (f) base-sixteen.
3. Write the number 413_{10} in (a) Egyptian numbers, (b) Roman numerals, (c) Babylonian numbers, (d) binary, and (e) base-sixteen.
4. Write the number $7,189_{10}$ in (a) Egyptian numbers, (b) Roman numerals, (c) Babylonian numbers, (d) binary, and (e) base-sixteen.
5. Write the number $23,724_{10}$ in (a) Egyptian numbers, (b) Roman numerals, (c) Babylonian numbers, (d) binary, and (e) base-sixteen.
6. Convert each of the tally-mark numbers in Fig. 1-23 into base-ten.

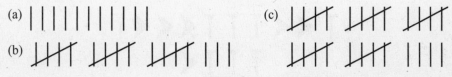

Fig. 1-23

7. Convert each of the Egyptian numbers in Fig. 1-24 into base-ten.

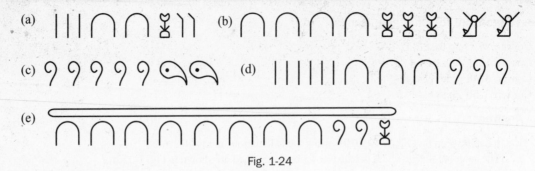

Fig. 1-24

8. Convert each of the Egyptian numbers in Fig. 1-24(a), 1-24(b), 1-24(c), and 1-24(d) into Roman numerals.
9. Convert each of these Roman numerals into base-ten: (a) XVII, (b) LXXIX, (c) CDXLIII, (d) MMMDCXX,
   (e) $\overline{\text{IV}}$CCIV, and (f) $\overline{\overline{\text{DLXXV}}}$XLIICCCXVI
10. Convert each of the Babylonian numbers in Fig. 1-25 into base-ten. Suppose that the x is placed to the right of the 1's place.

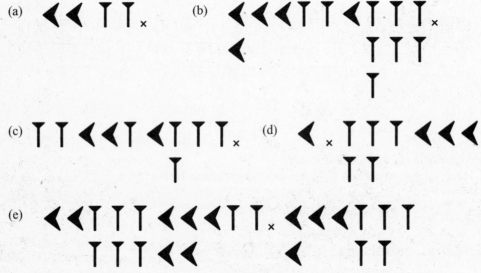

Fig. 1-25

11. Convert each of these binary numbers into base-ten: (a) 110_2, (b) 10011_2, (c) $1101\ 1100_2$, (d) $1101\ 1001.1101_2$, and (e) $10\ 0001\ 1001.001_2$.
12. Convert each of these hexadecimal numbers into base-ten: (a) D_{16}, (b) AF_{16}, (c) 125_{16}, (d) $30FF_{16}$, (e) $2E.04_{16}$, and (f) $275.B02_{16}$.
13. Convert each of these binary numbers into hexadecimal: (a) $1001\ 0101_2$, (b) $1100\ 0101\ 1110_2$, (c) $1\ 0111\ 0000\ 1110\ 1011_2$, and (d) $10\ 0001\ 1011\ 1011\ 0110\ 0101_2$.
14. Convert each of these hexadecimal numbers into binary: (a) $3B_{16}$, (b) 239_{16}, (c) $104A_{16}$, and (d) $BADC0DE_{16}$.
15. If the number system was unspecified, what might the number 111 mean?
16. Explain how the system of tally marks could be viewed as base-one.
17. Name five different numbers that the Babylonian number in Fig. 1-26 might represent.

Fig. 1-26

18. Suppose an ancient Babylonian letter uses the number in Fig. 1-27 to refer to a quantity of goats. How many goats, in base-ten, might this be if the letter were (a) a merchant requesting his goatherd nephew to sell some of the flock or (b) the great King Gilgamesh demanding tribute of a small city? Explain.

Fig. 1-27

Answers

1. (a) 1, 2, 3, 4, 5, 6, 7, 8, 9, 10, 11, 12, 13, 14, 15, 16, 17, 18, 19, 20, 21
   (b) Shown in Fig. 1-28.

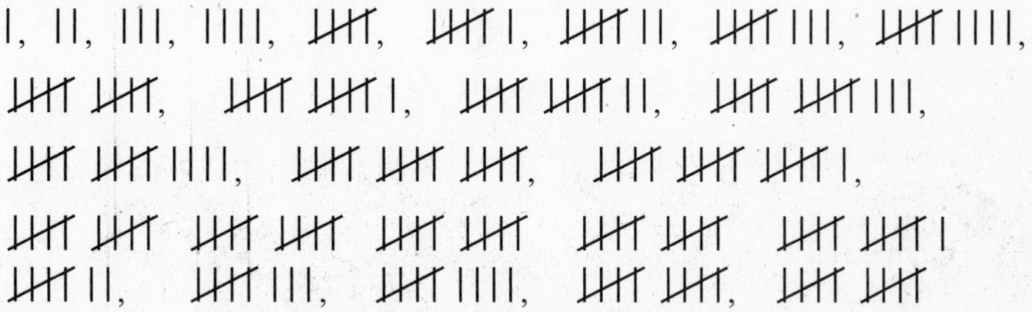

I, II, III, IIII, ⅢⅡ, ⅢⅡ I, ⅢⅡ II, Ⅲ III, Ⅲ IIII,
Ⅲ Ⅲ, Ⅲ Ⅲ I, Ⅲ Ⅲ II, Ⅲ Ⅲ III,
Ⅲ Ⅲ IIII, Ⅲ Ⅲ Ⅲ, Ⅲ Ⅲ Ⅲ I,
Ⅲ Ⅲ Ⅲ Ⅲ Ⅲ Ⅲ Ⅲ Ⅲ Ⅲ Ⅲ I
Ⅲ II, Ⅲ III, Ⅲ IIII, Ⅲ Ⅲ, Ⅲ Ⅲ

Fig. 1-28

   (c) Shown in Fig. 1-29.

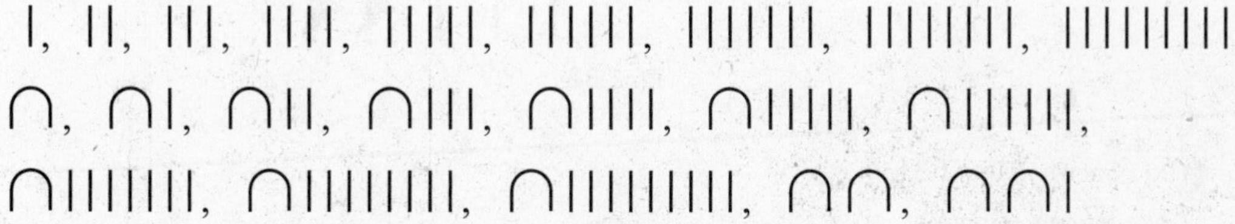

I, II, III, IIII, IIIII, IIIIII, IIIIIII, IIIIIIII, IIIIIIIII,
∩, ∩I, ∩II, ∩III, ∩IIII, ∩IIIII, ∩IIIIII,
∩IIIIIII, ∩IIIIIIII, ∩IIIIIIIII, ∩∩, ∩∩I

Fig. 1-29

   (d) I, II, III, IV, V, VI, VII, VIII, IX, X, XI, XII, XIII, XIV, XV, XVI, XVII, XVIII, XIX, XX, XXI
   (e) Shown in Fig. 1-30.

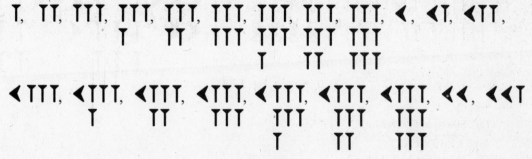

Fig. 1-30

   (f) 1, 10, 11, 100, 101, 110, 111, 1000, 1001, 1010, 1011, 1100, 1101, 1110, 1111, 10000, 10001, 10010, 10011, 10100, 10101
   (g) 1, 2, 3, 4, 5, 6, 7, 8, 9, A, B, C, D, E, F, 10, 11, 12, 13, 14, 15

2. (a) Shown in Fig. 1-31(a).
   (b) Shown in Fig. 1-31(b).
   (c) XLII
   (d) Shown in Fig. 1-31(d).
   (e) 10 1010
   (f) 2A

(a)

(b)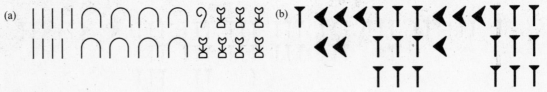

Fig. 1-31

3. (a) Shown in Fig. 1-32(a).
   (b) CDXIII
   (c) Shown in Fig. 1-32(b).
   (d) 1 1001 1101
   (e) 19D

Fig. 1-32

4. (a) Shown in Fig. 1-33(a).
   (b) $\overline{\text{VII}}$CLXXXIX
   (c) Shown in Fig. 1-33(b).
   (d) 1 1100 0001 0101
   (e) 1C15

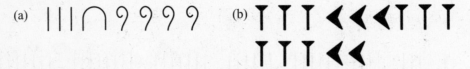

Fig. 1-33

5. (a) Shown in Fig. 1-34(a).
   (b) $\overline{\text{XXIII}}$DCCXXIV
   (c) Shown in Fig. 1-34(b).
   (d) 101 1100 1010 1100
   (e) 5CAC

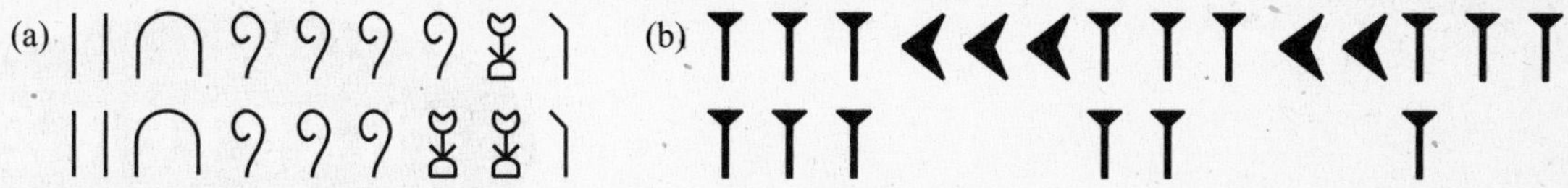

(a)

(b)

Fig. 1-34

6. (a) 11, (b) 18, and (c) 29

7. (a) 21,023, (b) 2,013,040, (c) 200,500, (d) 336, and (e) $\dfrac{1}{1,280}$

8. (a) $\overline{\text{XXIXXIII}}$, (b) $\overline{\overline{\text{II}}}\text{XXIIIXL}$, (c) $\overline{\text{CCD}}$, and (d) CCCXXXVI

9. (a) 17, (b) 79, (c) 443, (d) 3,620, (e) 4,204, and (f) 575,042,316

10. (a) 22, (b) 2,537, (c) 8,474, (d) $10 + \dfrac{5}{60} + \dfrac{30}{3600} = 10.09\overline{16}$, and (e) 1,612.75

11. (a) 6, (b) 19, (c) 220, (d) 217.8125, and (e) 537.125

12. (a) 13, (b) 175, (c) 293, (d) 12,543, (e) 46.015625, and (f) $629 + \dfrac{11}{16} + \dfrac{2}{4,096} \approx 629.687988$

13. (a) 95_{16}, (b) $C5E_{16}$, (c) $170EB_{16}$, and (d) $21BB65_{16}$

14. (a) $11\ 1011_2$, (b) $10\ 0011\ 1001_2$, (c) $1\ 0000\ 0100\ 1010_2$, and (d) $1011\ 1010\ 1101\ 1100\ 0000\ 1101\ 1110_2$

15. 3 in tally marks, 7 in binary, 111 in base-ten, and 273 in hexadecimal

16. A base-one numbering system would employ only one symbol. Each position would represent a power of 1. Because 1 to any power is 1, this means that each of the symbols has the same value: 1. This describes the tally mark system completely.

17. If 1's place is rightmost, this number represents 5,201. Switching the 1's place changes the number by a power of 60; thus, the number could represent 5,201, 312,060, 18,723,600, $\dfrac{5,201}{60}$, $\dfrac{5,201}{3,600}$, etc.

18. Because fractions of goats make little sense, this number represents 80, 4,800, 288,000, 17,280,000, etc. In (a) the number is likely 80, whereas in (b) the number is likely 4,800 unless the city has many hundreds of thousands of goats. Gilgamesh would not bother to write a letter for only 80 goats.

CHAPTER 2

Sets

Sets

A *set* is a collection of objects called *elements*. For example, an alphabet is a set with letters for elements.

A capital letter is often used to represent a set. For example, let A=the English alphabet. We write $d \in A$ to indicate that the letter d is an element of set A. We write $7 \notin A$ to say that the number 7 is not an element of A.

Curly brackets { and } are used to indicate sets. For example, $B = \{2, 3, 5, 7\}$ is a set with four elements: 2, 3, 5, and 7.

Large sets are often described with ellipses. These require recognition of the pattern. For example, the natural numbers are $\mathbf{N} = \{1, 2, 3, ...\}$ and the integers are $\mathbf{Z} = \{... -3, -2, -1, 0, 1, 2, 3...\}$ (Z is for *Zahl*, the German word for *number*).

Other common symbols are $\mathbf{Q}$=the set of all rational numbers (quotients of integers), $\mathbf{R}$=the set of all real numbers, and $\mathbf{C}$=the set of all complex numbers. For example, $x \in \mathbf{R}$ means that the letter x represents a real number, and $\sqrt{2} \notin \mathbf{Q}$ means that the square root of 2 is not rational.

It is also common to describe a set with *set-builder notation*: $\{x \mid P(x)\}$. Within the curly brackets is a vertical line. To the left is a variable x, which will represent an element of the set. To the right of the vertical line are the requirements $P(x)$ for membership in the set.

For example, $R^+ = \{x \in \mathbf{R} \mid x > 0\}$ represents the set of all real numbers that are greater than zero. The set $D = \{n \in \mathbf{N} \mid n \text{ is prime and } n < 10\}$ consists of all the natural numbers that are both prime and less than 10, namely, 2, 3, 5, and 7. This makes D equal to the set B defined earlier. Two sets are *equal* if they have the exact same elements.

Only the elements define a set, not the manner or order in which they are presented. For example, sets $E = \{1, 2, 3\}$ and $F = \{3, 2, 1\}$ are equal because they contain the same three elements.

A set with no elements is called the *null set* or the *empty set* and is written $\emptyset$. There is only one null set because any other set with no elements will have the same elements: none at all.

SOLVED PROBLEMS

Elements of Sets

1. Let sets $A = \{2, 4, 6, 8, 10\}$, $B = \{3, 6, 9, 12,...\}$, and $C = \{x \in \mathbf{R} \mid 0 \leq x \leq 100\}$. Answer true or false: (a) $1 \in A$, (b) $2 \in A$, (c) $30 \in B$, (d) $25 \in C$, (e) $\sqrt{5} \in C$, (f) $\sqrt{36} \notin B$, (g) $6 \in \emptyset$, (h) $5 \in \mathbf{Q}$, and (i) $\sqrt{5} \in \mathbf{Q}$.

2. Write out all the elements of (a) $D = \{x \in \mathbf{Z} \mid -3 \leq x < 5\}$, (b) $E = \{y \in \mathbf{R} \mid y^2 = 100\}$, and (c) $F = \{n \in \mathbf{N} \mid n + 5 < 12\}$.

3. Yes or no:

   (a) $\{1, 2, 3, 4\} = \{4, 3, 2, 7\}$

   (b) $\{a, b, 1\} = \{1, a, b\}$

   (c) $\{3, 5, 3, 7\} = \{7, 5, 3\}$

(d) $\{y \in \mathbf{N}\,|\,y \text{ is even}\} = \{2, 4, 6, 8,...\}$

(e) $\mathbf{N} = \{x \in \mathbf{Z}\,|\,x \geq 0\}$

(f) $\mathbf{Q} = \left\{\dfrac{a}{b}\,\middle|\,a \in \mathbf{Z}, b \in \mathbf{N}\right\}$

Answers

1.   (a) $1 \in A$ is false because 1 is not an element of A.

 (b) $2 \in A$ is true because 2 is an element of A.

 (c)   It looks like B consists of all positive multiples of 3, so $30 \in B$ is true.

 (d) $25 \in C$ is true because $25 \in \mathbf{R}$ and $0 \leq 25 \leq 100$.

 (e) $\sqrt{5} \in C$ is also true because $\sqrt{5}$ is real, positive, and less than 100.

 (f) $\sqrt{36} \notin B$ is false because $\sqrt{36} = 6$ is an element of B.

 (g) $6 \in \varnothing$ is false because $\varnothing$ has no elements.

 (h) $5 \in \mathbf{Q}$ is true because $5 = \dfrac{5}{1}$ is rational.

 (i) $\sqrt{5} \in \mathbf{Q}$ is false because 5 is not a perfect square, so $\sqrt{5}$ is *irrational* (not rational).

2.   (a) $D = \{-3, -2, -1, 0, 1, 2, 3, 4\}$

 (b) $E = \{10, -10\}$

 (c) $F = \{1, 2, 3, 4, 5, 6\}$

3.   (a)   No: only one contains 7.

 (b)   Yes.

 (c)   Yes: the first set only has three different elements.

 (d)   Yes.

 (e)   No, but only because 0 is in the second set.

 (f)   Yes, any rational number is an integer divided by a natural number.

Subsets

Set A is a *subset* of set B if all the elements of A are elements of B. This is written $A \subset B$.

For example, if $B = \{1, 2, 3,..., 10\}$ and $A = \{3, 4, 5\}$, then $A \subset B$ because the three elements of A are also elements of B. If $C = \{0, 1, 2\}$, then $C \not\subset B$ because $0 \in C$ but $0 \notin B$.

A set is technically a subset of itself because all of its elements are in the set. Thus, $A \subset A$ for every set A.

Technically, $\varnothing$ is a subset of every set because all of its elements (all zero of them) are in every other set. Thus, $\varnothing \subset A$ for every set A.

While anything can be an element, only sets can be subsets. For example, $1 \in \{1, 2, 3\}$, but $1 \not\subset \{1, 2, 3\}$ because the number 1 is not a set. The subsets of $\{1, 2, 3\}$ are $\{1\}$, $\{2\}$, $\{3\}$, $\{1, 2\}$, $\{1, 3\}$, $\{2, 3\}$, $\{1, 2, 3\}$, and $\varnothing$. Notice that 1 and $\{1\}$ are different objects; the first is a number (a quantity) and the second is a set (a collection).

The set of all subsets of a set A is called the *power set* and is written $P(A)$. For example, $P(\{1, b\}) = \{\{1\}, \{b\}, \{1, b\}, \varnothing\}$.

The set of all elements under discussion is the *universal set*, usually written U. For example, when discussing the students of a class, the universal set might be the set of all students in the class.

SOLVED PROBLEMS

Subsets

1. True or false:

    (a) $\{1, 4, 5\} \subset \{1, 2, 3, 4, 5\}$
    (b) $\{1, 2, 3\} \subset \{3\}$
    (c) $\{1, 2\} \subset \{1, 3, 5\}$
    (d) $\{1\} \not\subset \{1, 2, 3, \dots, 100\}$
    (e) $\varnothing \subset \{1, 2, 3, 4, 5\}$
    (f) $\{x \mid x \text{ is prime}\} \subset \mathbf{N}$
    (g) $\mathbf{N} \subset \mathbf{R}$
    (h) $\mathbf{R} \subset \mathbf{C}$
    (i) $\mathbf{Z} \subset \mathbf{N}$
    (j) $\mathbf{Q} \subset \mathbf{C}$

2. Name the power set of the following:

    (a) $\{a, b\}$
    (b) $\varnothing$
    (c) $\{1, 2, 3, 4\}$

    #### Answers

    1. (a) True.
        (b) False: 1 is an element of the first set but not the second.
        (c) False: 2 is an element of the first set but not the second.
        (d) False: $\{1\}$ is a subset of $\{1, 2, 3, \dots, 100\}$.
        (e) True: $\varnothing$ is a subset of every set.
        (f) True: all prime numbers are natural.
        (g) True: all natural numbers are real.
        (h) True: all real numbers are complex numbers with imaginary part 0.
        (i) False: $-1 \in \mathbf{Z}$ but $-1 \notin \mathbf{N}$.
        (j) True: rational numbers are all real, thus also complex.

    2. (a) $P(\{a,b\}) = \{\{a\}, \{b\}, \{a, b\}, \varnothing\}$
        (b) $P(\varnothing) = \{\varnothing\}$ because $\varnothing$ is the only subset of $\varnothing$
        (c) $P(\{1, 2, 3, 4\}) = \{\{1\}, \{2\}, \{3\}, \{4\}, \{1, 2\}, \{1, 3\}, \{1, 4\}, \{2, 3\}, \{2, 4\}, \{3, 4\}, \{1, 2, 3\}, \{1, 2, 4\}, \{1, 3, 4\}, \{2, 3, 4\}, \{1, 2, 3, 4\}, \varnothing\}$

Unions and Intersections

The *union* of two sets is a new set with all the elements of both sets combined. The union of sets A and B is written $A \cup B$. For example, if $A = \{1, 2, 3, 4\}$ and $B = \{2, 4, 6, 8, 10\}$, then $A \cup B = \{1, 2, 3, 4, 6, 8, 10\}$.

The *intersection* of two sets is a new set containing only those elements that are in both sets. The intersection of A and B is written $A \cap B$. For example, $\{1, 2, 3, 4\} \cap \{2, 4, 6, 8, 10\} = \{2, 4\}$.

Try Fig. 2-1 for a mnemonic (memory aid) to union and intersection.

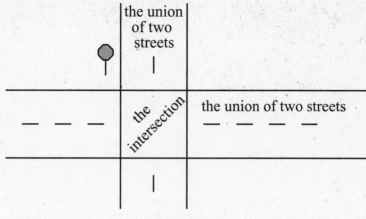

Fig. 2-1

Union and intersection are commutative and associative operations. Thus, for any sets A, B, and C, $A \cup B = B \cup A$, $A \cap B = B \cap A$, $(A \cup B) \cup C = A \cup (B \cup C)$, and $(A \cap B) \cap C = A \cap (B \cap C)$.

Furthermore, $A \cap (B \cup C) = (A \cap B) \cup (A \cap C)$ and $A \cup (B \cap C) = (A \cup B) \cap (A \cup C)$.

Also, $A \subset A \cup B$, $A \cap B \subset A$, $\varnothing \cap A = \varnothing$, and $A \cup \varnothing = A$.

Two sets are called *disjoint* if their intersection is the null set.

SOLVED PROBLEMS

Union and Intersection

Write out the sets formed by the following:

1. $\{a, b, c, d\} \cup \{d, e, f\}$
2. $\{1, 2, a\} \cup \{1, x, y\}$
3. $\{1, 2, 3, \ldots, 10\} \cup \{100, 101, \ldots, 110\}$
4. $\{1, 2, \ldots, 10\} \cup \{3, 4, 7\}$
5. $\{a, b, c, d\} \cap \{d, e, f\}$
6. $\{1, 2, 3, \ldots, 10\} \cap \{100, 101, \ldots, 110\}$
7. $\{2, 3, 5, 7, 11, 13\} \cap \{1, 2, 3, 4, 5\}$
8. $\{1, 2, \ldots, 10\} \cap \{3, 4, 7\}$
9. $\{a, b, c, d, e\} \cup \varnothing$
10. $\varnothing \cap \{1, 2, 3\}$
11. $\{3, 5, 10, 14\} \cup \{5, 10, 15, 20\} \cup \{1, 2, 4, 8\}$
12. $\{1, 2, 3, \ldots, 10\} \cap \{5, 6, 7, \ldots, 20\} \cap \{2, 4, 6, 8, 10\}$
13. $\mathbf{R} \cap \mathbf{Q}$
14. $\mathbf{Z} \cup \mathbf{N}$
15. $(\{1, 2, 3, 4, 5\} \cup \{3, 6, 9\}) \cap \{2, 4, 6, 8, 10\}$

Answers

1. $\{a, b, c, d, e, f\}$
2. $\{1, 2, a, x, y\}$
3. $\{1, 2, 3, 4, 5, 6, 7, 8, 9, 10, 100, 101, 102, 103, 104, 105, 106, 107, 108, 109, 110\}$
4. $\{1, 2, 3, 4, 5, 6, 7, 8, 9, 10\}$
5. $\{d\}$
6. $\varnothing$
7. $\{2, 3, 5\}$
8. $\{3, 4, 7\}$

9. {a, b, c, d, e}
10. ∅
11. {1, 2, 3, 4, 5, 8, 10, 14, 15, 20}
12. {6, 8, 10}
13. **Q**
14. **Z**
15. {2, 4, 6}

Venn Diagrams

Relationships between sets can be illustrated by pictures called *Venn diagrams*. Each set is represented by a closed figure, usually a circle or an oval labeled inside by the set's letter. An element is illustrated by a labeled dot. A set's subsets are drawn in its interior. A big rectangle around everything represents the universal set.

For example, $A \subset B$ and $B \subset C$ with universal set U is illustrated in Fig. 2-2. Here, $x \in B$ and $x \in C$ but $x \notin A$.

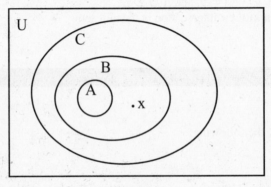

Fig. 2-2

Disjoint sets are illustrated by nonoverlapping figures. For example, $A \cap B = \emptyset$ is illustrated in Fig. 2-3. A point could be drawn in A or in B, but not in both.

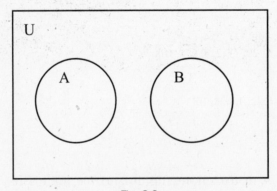

Fig. 2-3

A Venn diagram must illustrate the most general possible situation. For two sets A and B, with no further information about their relationship, this looks like Fig. 2-4. Elements could be in both A and B, in A but not B, in B but not A, or in neither A nor B (for example, z in the figure).

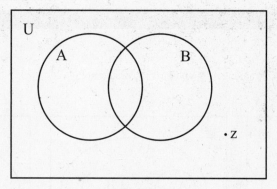

Fig. 2-4

The set of elements in A but not in B is written $A - B$. The set of all elements not in A (but in the universal set) is called the *complement* of A, and written either $U - A$ or A^c.

For three sets A, B, and C, the most general Venn diagram is in Fig. 2-5.

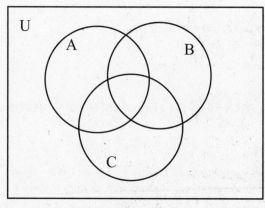

Fig. 2-5

One can verify that $A \cap (B \cup C) = (A \cap B) \cup (A \cap C)$ by seeing that both sets are represented by the shaded region in Fig. 2-6.

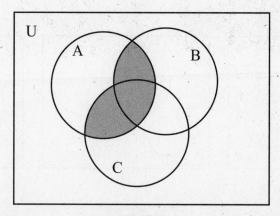

Fig. 2-6

SOLVED PROBLEMS

Venn Diagrams

1. Name all the set relationships among *A*, *B*, and *C* as illustrated in Fig. 2-7.

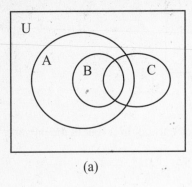
(a)

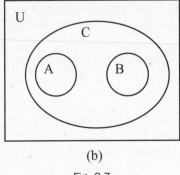
(b)

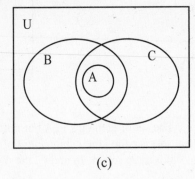
(c)

Fig. 2-7

2. Illustrate the most general Venn diagram with the following:

   (a) $A \subset B$ and $A \cap C = \varnothing$
   (b) $A \cap B = \varnothing$ and $A \cap C = \varnothing$
   (c) $C \subset B$ and $B \cap A = \varnothing$

3. Draw a Venn diagram for the sets M = the set of all mammals, H = the set of all humans, W = the set of all women, and L = the set of all lawyers.

4. Suppose $U = \{1, 2, 3, 4, 5, 6, 7, 8, 9, 10\}$, $A = \{2, 4, 6, 8\}$, $B = \{3, 6, 9\}$, and $C = \{1, 5, 10\}$. What are (a) $A - B$, (b) $B - A$, (c) $C - A$, (d) $A - U$, (e) A^c, and (f) C^c?

5. Suppose there are 100 seniors in a high school. Eighty of them are taking physics, and 60 of them play sports. What are all the possibilities for the number of students who take physics and also play sports?

   ### Answers

   1. (a) $B \subset A$
      (b) $A \subset C$, $B \subset C$, and $A \cap B = \varnothing$
      (c) $A \subset B$ and $A \subset C$

   2. See Fig. 2-8.

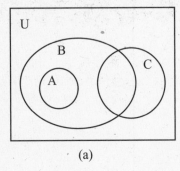
(a)

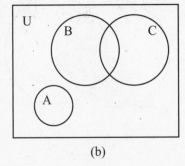
(b)

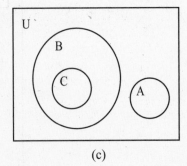
(c)

Fig. 2-8

3. The key details here are that M is the universal set (all elements are mammals), all women and lawyers are human, and some lawyers are women. This is illustrated in Fig. 2-9.

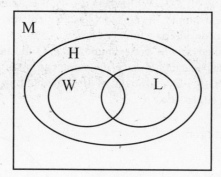

Fig. 2-9

4. (a) $A-B=\{2, 4, 8\}$, (b) $B-A=\{3, 9\}$, (c) $C-A=\{1, 5, 10\}$, (d) $A-U=\varnothing$, (e) $A^c=\{1, 3, 5, 7, 9, 10\}$, and (f) $C^c=\{2, 3, 4, 6, 7, 8, 9\}$

5. It is possible that all the physics students play sports. Thus, 60 is the largest number of students who both take physics and play sports. It is not possible that the set of physics students is disjoint with the set of athletes, because that would require at least $60+80=140$ people in our universal set of 100. The least situation occurs when all 20 of the people who do not study physics are athletes. The remaining 40 must thus take physics, so 40 is the least number of students who take physics and play sports. Any number from 40 to 60 could occur.

Russell's Paradox

There was a time when mathematicians used sets like $U=$ the set of all sets. Because U was considered a set, this meant that U was an element of itself: $U \in U$. This was curious because most sets are not elements of themselves. For example, $\{1, 2, 3\} \notin \{1, 2, 3\}$. How could you assemble a collection of things if one of the ingredients was the finished collection?

In 1901, Bertrand Russell found a contradiction in this sort of set theory. He defined a set $S=\{A \in U \,|\, A \notin A\}$, the collection of all sets that are not elements of themselves. For example, $\{1, 2, 3\} \in S$, but $U \notin S$ because $U \in U$.

Russell then asked: is $S \in S$? It is always a reasonable question to check whether something is an element of a set. We expect that either S is an element of itself like U, or not, like $\{1, 2, 3\}$.

If $S \in S$, then S is an element of $S=\{A \in U \,|\, A \notin A\}$, so S is one of those sets which are not in themselves: $S \notin S$.

If $S \notin S$, then S satisfies the one requirement to be an element of S: it is not an element of itself. Thus, $S \in S$.

In either case, we have a contradiction. This is called *Russell's paradox*. To escape contradiction, we must find where the problem began. In this case, we recognize that S is an impossible object. This teaches us that we cannot always use set-builder notation to define a set. (There are also other ways to avoid the paradox.) Russell's paradox suggests that the collection of all things may be too big to be a thing itself. Some mathematicians distinguish arbitrary collections (called *classes*) from those collections that do not lead to contradictions (and these collections are called *sets*).

Cantor's Diagonal Proof

Two sets have the same *cardinality* if their elements can be entirely paired up. For example, $A=\{1, 3, 5\}$ and $B=\{2, a, b\}$ have the same cardinality because we can pair 1 with 2, 3 with a, and 5 with b. Pairs are sometimes represented with sets of parentheses, for example (1, 2), (3, a), and (5, b). The set $C=\{x, y\}$ has a different (smaller) cardinality because it has too few elements to pair completely with A or B.

A set with the same cardinality as $\{1, 2, 3,…, n\}$, where $n \in \mathbf{N}$, is called a *finite* set. Here, n represents the number of elements in the set. For example, the alphabet has 26 elements, the same cardinality as set $\{1, 2, 3, 4,…, 26\}$. The empty set is also considered a finite set.

Sets which are not finite are called *infinite*. For example, $\mathbf{N}, \mathbf{Z}, \mathbf{Q}, \mathbf{R}$, and $\mathbf{C}$ are all infinite sets.

The natural numbers $\mathbf{N}$ and the integers $\mathbf{Z}$ have the same cardinality because their elements can be paired: (1, 0), (2, 1), (3, −1), (4, 2), (5, −2), and so on, as illustrated in Fig. 2-10. This entirely pairs their elements: no element is missing on either side.

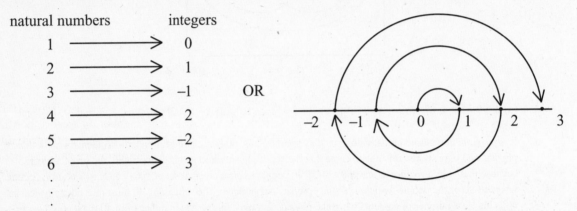

Fig. 2-10

The natural numbers and the rational numbers $\mathbf{Q}$ have the same cardinality. If all the rational numbers are laid out in rows with common (reduced) denominators, then we can zigzag our way through them all, counting along with the natural numbers, as shown in Fig. 2-11.

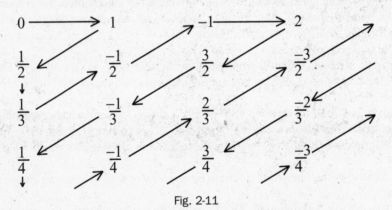

Fig. 2-11

This pairing begins as follows: $(1,0),\ (2,1),\ \left(3,\dfrac{1}{2}\right),\ \left(4,\dfrac{1}{3}\right),\ \left(5,-\dfrac{1}{2}\right),\ (6,-1),\ (7,2),\ …$

The natural numbers $\mathbf{N}$ and the real numbers $\mathbf{R}$ have different cardinality. It is impossible to pair up the natural numbers with the real numbers, even though both sets are infinite.

Suppose someone tried to pair up $\mathbf{N}$ with $\mathbf{R}$. Some irrational numbers can only be described with an infinite string of decimal digits, so we can suppose that all the real numbers are written out to infinity, with zeros if necessary. An example is illustrated in Fig. 2-12.

Cantor's diagonal proof explains why this cannot be a complete pairing between $\mathbf{N}$ and $\mathbf{R}$. Look at the "diagonal" formed by the first digit after the decimal place of the first number, the second digit of the second number, and so on, as illustrated in Fig. 2-13. In this example, the diagonal begins 0.30450 . . . If we change every digit of this string, we will get a real number that is not on the list in the example. For example,

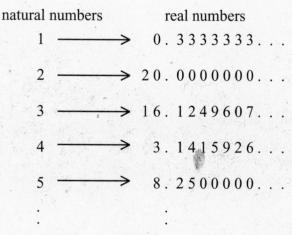

natural numbers   real numbers

1 ⟶ 0 . 3 3 3 3 3 3 3 . . .

2 ⟶ 2 0 . 0 0 0 0 0 0 0 . . .

3 ⟶ 1 6 . 1 2 4 9 6 0 7 . . .

4 ⟶ 3 . 1 4 1 5 9 2 6 . . .

5 ⟶ 8 . 2 5 0 0 0 0 0 . . .

Fig. 2-12

0.55545 . . . is not the first number on the list because the first digit after its decimal place is different from the first decimal digit of that number. Similarly, 0.55545 . . . is not the second number on the list, nor the third, etc. For that matter, 0.21788 . . . and 0.11177 . . . are not anywhere on the list, as long as they are different from the diagonal in every place. Because there are real numbers missing from this pairing, it is not a complete pairing. We conclude that **N** and **R** must have different cardinalities. There are simply too many more real numbers than there are natural numbers.

all real numbers?

0 . ③ 3 3 3 3 3 3 . . .

2 0 . 0 ⓪ 0 0 0 0 0 . . .

1 6 . 1 2 ④ 9 6 0 7 . . .

3 . 1 4 1 ⑤ 9 2 6 . . .

8 . 2 5 0 0 ⓪ 0 0 . . .

Fig. 2-13

We were able to list out all the rational numbers, but cannot list out the real numbers, so it must be the irrational numbers which are too numerous to count. Thus, most of the real numbers are of the annoying sort, which take forever to write out exactly. Only a very few of them can be described by a fraction.

In short, there are different kinds of infinity.

SUPPLEMENTAL PROBLEMS

1.   Let $A=\{1, 2, 3, 4, 5\}$, $B=\{2, 4, 6, 8,...\}$, and $C=\{x\in\mathbf{Q}\,|\,x>5\}$. State true or false:

(a) $10\in A$

(b) $100\in B$

(c) $4\in C$

(d) $\sqrt{30}\in C$

(e) $\sqrt{81} \in C$

(f) $5 \in \varnothing$

(g) $8 \in \mathbf{N}$

(h) $-3 \notin \mathbf{N}$

(i) $\sqrt{5} \in \mathbf{R}$

2. Write out all the elements of the following:

(a) $A = \{x \in \mathbf{N} \mid x \text{ is odd and } x < 12\}$

(b) $B = \{x \in \mathbf{Z} \mid \text{the absolute value } |x| < 4\}$

(c) $C = \{x \in \mathbf{R} \mid x^2 = 5\}$

3. State true or false:

(a) $\{a, b, c, d, f\} = \{a, b, c, e, f\}$

(b) $\{2, 3, 5, 7, 1\} = \{7, 5, 1, 2, 3\}$

(c) $\{5, 10, 5, 15, 5, 20\} = \{5, 10, 15, 20\}$

(d) $\{1, 2, 3\} = \{a, b, c\}$

(e) $\mathbf{N} = \{x \in \mathbf{Z} \mid x > 0\}$

(f) $\{x \in \mathbf{R} \mid x^2 = x\} = \{0, 1\}$

4. State true or false:

(a) $\{42, 99\} \subset \{1, 2, 3, \ldots, 100\}$

(b) $\{-2, 4, 7, 12\} \subset \{-3, -2, 0, 1, 3, 4, 6, 7, 8, 12\}$

(c) $\{6, 7, 8\} \subset \{2, 4, 6, 8\}$

(d) $\{1, 2\} \subset \varnothing$

(e) $\varnothing \subset \{3, 6, 9, \ldots\}$

(f) $\{5, 10, 15, 20, \ldots\} \subset \mathbf{Z}$

(g) $\mathbf{Z} \subset \mathbf{R}$

(h) $\mathbf{Q} \subset \mathbf{N}$

5. Name the power set of the following:

(a) $\{3, 5\}$

(b) $\{7\}$

(c) $\{x, y, z\}$

6. Write out the set formed by the following:

(a) $\{3, 6, 9\} \cup \{1, 2, 3, 4, 5\}$

(b) $\{2, 4\} \cup \{a, b, c\}$

(c) $\{1, 2, 3, \ldots 10\} \cup \{3, 8, 9\}$

(d) $\{5, 10, 15, 20\} \cap \{1, 2, 3, \ldots, 10\}$

(e) $\{a, b, c, d, e, f\} \cap \{b, e, f\}$

(f) $\{-5, -3, -2, -1\} \cap \{3, 4, 5, 7\}$

(g) $\{3, 8, 11, 13\} \cup \varnothing$

(h) $\{3, 8, 11, 13\} \cap \varnothing$

(i) $\{1, 2, 3, 4, 5\} \cup \{3, 4, 5, 6, 7, 8, 9, 10\} \cup \{2, 4, 6, 8, 10, 12, 14\}$

(j) $\{2, 4, 6, 8, \ldots, 20\} \cap \{3, 6, 9, 12, 15, 18\} \cap \{1, 2, 3, \ldots, 10\}$

(k) $(\{3, 4, 7, 8, 9\} \cap \{2, 4, 6, 8\}) \cup \{1, 2, 3, 4, 5\}$

(l) $\mathbf{Q} \cap \mathbf{Z}$

(m) $\mathbf{R} \cup \mathbf{N}$

7. Suppose $A = \{a, b, c\}$, $B = \{d, e, f\}$, and $C = \{a, c, d\}$, and the universal set is $U = \{a, b, c, d, e, f, g\}$. Find the following.

(a) $A - B$

(b) $A - C$

    (c) $B - C$

    (d) $C - A$

    (e) $C - C$

    (f) A^c

    (g) C^c

8. Name all the relationships between sets A, B, and C illustrated in Fig. 2-14.

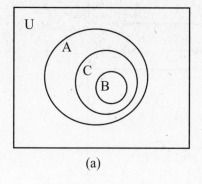 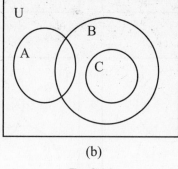 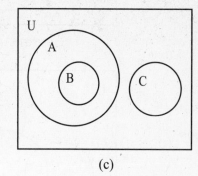

       (a)   (b)   (c)

Fig. 2-14

9. Illustrate the most general Venn diagram for sets A, B, and C, where the following apply.

    (a) $A \cap B = \varnothing$ and $B \cap C = \varnothing$

    (b) $C \subset A$ and $C \subset B$

    (c) $A \cap B = \varnothing$

    (d) $A \cap B \cap C = \varnothing$

10. Draw a Venn diagram for the sets $A =$ the set of all animals, $B =$ the set of all bears, $C =$ the set of all circus animals, and $D =$ the set of all dogs.

11. Suppose that in a room of 42 people, 35 speak English and 22 speak Spanish. What are the possible numbers of people who speak both English and Spanish?

Answers

1.   (a) false, (b) true, (c) false, (d) false, (e) true, (f) false, (g) true, (h) true, and (i) true

2.   (a) $A = \{1, 3, 5, 7, 9, 11\}$, (b) $B = \{-3, -2, -1, 0, 1, 2, 3\}$, and (c) $C = \{-\sqrt{5}, \sqrt{5}\}$

3.   (a) false, (b) true, (c) true, (d) false, (e) true, and (f) true

4.   (a) true, (b) true, (c) false, (d) false, (e) true, (f) true, (g) true, and (h) false

5.   (a) $P(\{3, 5\}) = \{\{3\}, \{5\}, \{3, 5\}, \varnothing\}$

    (b) $P(\{7\}) = \{\{7\}, \varnothing\}$

    (c) $P(\{x, y, z\}) = \{\{x\}, \{y\}, \{z\}, \{x, y\}, \{x, z\}, \{y, z\}, \{x, y, z\}, \varnothing\}$

6.   (a) $\{1, 2, 3, 4, 5, 6, 9\}$, (b) $\{2, 4, a, b, c\}$, (c) $\{1, 2, 3, 4, 5, 6, 7, 8, 9, 10\}$, (d) $\{5, 10\}$, (e) $\{b, e, f\}$, (f) $\varnothing$, (g) $\{3, 8, 11, 13\}$, (h) $\varnothing$, (i) $\{1, 2, 3, 4, 5, 6, 7, 8, 9, 10, 12, 14\}$, (j) $\{6\}$, (k) $\{1, 2, 3, 4, 5, 8\}$, (l) $\mathbf{Z}$, and (m) $\mathbf{R}$

7.   (a) $\{a, b, c\}$, (b) $\{b\}$, (c) $\{e, f\}$, (d) $\{d\}$, (e) $\varnothing$, (f) $\{d, e, f, g\}$, and (g) $\{b, e, f, g\}$

8.   (a) $B \subset C$, $B \subset A$, and $C \subset A$

    (b) $C \subset B$ and $A \cap C = \varnothing$

    (c) $B \subset A$, $A \cap C = \varnothing$, and $B \cap C = \varnothing$

9.   See Fig. 2-15.

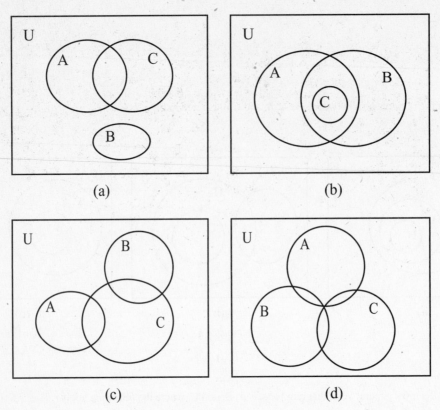

(a)   (b)

(c)   (d)

Fig. 2-15

10. All bears, dogs, and circus animals are animals. No animal is both a bear and a dog. Some dogs and some bears are circus animals. See Fig. 2-16.

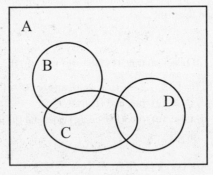

Fig. 2-16

11. As few as 15 and as many as 22 people speak both English and Spanish.

CHAPTER 3

Logic

Logic studies the manner in which consequences can be drawn: how truths can be combined to form new truths.

Definitions

One key to the success of mathematics is the precise nature of its definitions. Every word used either is found in ordinary dictionaries or else is defined somewhere in the book, usually on the first page listed for the word in the book's index.

It has been said that every word can be defined in 25 words or less. For example, a *tree* is a large perennial plant with one trunk and many branches (13 words). There is no need to name the different kinds or uses of trees.

Curiously, the most basic words are the hardest to define. For example, it is difficult to spell out what makes a line *straight* or not. In geometry, a *point* is often viewed as a location in space and *space* is often considered as the collection of all points. Circular definitions like this are unsatisfactory. In geometry, words like *point*, *line*, and *plane* are often defined only loosely because there are no more basic words with which to describe them.

SOLVED PROBLEMS

Definitions

1. Use this book's index to find the definition of the following:

    (a) a theorem
    (b) the mean
    (c) an annuity
    (d) irrational
    (e) a fractal
    (f) a permutation
    (g) an angle
    (h) a polygon

2. Define the following terms in 25 words or less:

    (a) car
    (b) dog
    (c) cat
    (d) job
    (e) table

Answers

1.   (a) A *theorem* is a mathematical statement that can be proven true.
     (b) The *mean* of a list of numbers is obtained by adding up all the numbers and dividing by the number of data entries.
     (c) An *annuity* is a positive amount of money from which regular withdrawals are made.
     (d) An *irrational* number is not rational (a ratio of integers).
     (e) A *fractal* is an object which properly contains things which resemble the entire object.
     (f) A *permutation* is a way of putting different objects in order.
     (g) An *angle* is formed when two lines come together at a point.
     (h) A *polygon* is a figure enclosed by straight sides.

2.   (a) A *car* is an enclosed four-wheeled motorized vehicle capable of transporting people for long distances.
     (b) A *dog* is a domestic companion animal distantly related to the wolf.
     (c) A *cat* is a domestic companion animal distantly related to lions and tigers. (Note that without mentioning ancestry, it would be hard to differentiate between cats and dogs.)
     (d) A *job* is an agreement to work over a period of time in exchange for payment.
     (e) A *table* is a raised, flat, and level surface.

Statements

The building blocks of logic are *statements*: sentences which must be either true or false. For example, "Mercury is more dense than copper" is a statement. We might not know if it is true or false, but the sentence leaves no third possibility. A sentence like "You are 6 feet tall" is not a statement unless it is very clear to whom "you" refers. Opinions like "Paris is beautiful" are best avoided in science, logic, and mathematics because they can be true for some people and false for others.

Sentences which refer to themselves can involve paradoxes. For example, "This sentence is false" involves a paradox very similar to Russell's paradox. If this sentence is true, then it is false. If the sentence is false, then it is true. This sentence is not exactly true or false, and thus is not the sort of statement used in logic.

The easiest logical consequence is that the exact opposite of a statement, its *negation*, is false when the original is true and true when the original is false. For example, the negation of "Mercury is more dense than copper" is "Mercury is not more dense than copper." One of these statements is true, and the other must be false.

To simplify things, statements can be represented by capital letters. The negation of a statement P is written $\neg P$.

In general, the negation of an "all," "every," or "none" statement will involve "There is at least one . . ." For example, if M = "All mammals require sleep," then $\neg M$ = "There is at least one mammal that does not require sleep."

The negation of a statement with "some" will involve some variety of "none." For example, the negation of "Some children fear the dark" is "No child fears the dark."

SOLVED PROBLEMS

Statements

Identify the following sentences as statements or not statements. Give the negation of each statement.

(a) The moon is full tonight.
(b) All ninjas hate pirates.
(c) Some pirates love other pirates.

(d) Granite is a metamorphic rock.

(e) The cost is unimportant.

(f) Buy me a milkshake.

(g) All horses are mammals.

(h) There are 12 kinds of amino acid.

(i) Dostoevsky was a genius.

Answers

(a) This is a statement, with negation "The moon is not full tonight."

(b) This is a statement, with negation "At least one ninja does not hate pirates."

(c) This is a statement (assuming that "love" is well defined), with negation "No pirate loves other pirates."

(d) This is a statement, with negation "Granite is not a metamorphic rock."

(e) This is not a statement because it is not clear to what "the cost" refers and because "unimportant" is an opinion.

(f) This is a command and not a true or false statement.

(g) This is a statement, with negation "At least one horse is not a mammal."

(h) This is a statement, with negation "There are not exactly 12 kinds of amino acid."

(i) This is an opinion and not a statement.

Conjunctions

Statements can be combined by the conjunctions *and* and *or* to form compound statements.

Suppose P and Q are statements. The statement "P and Q," written $P \wedge Q$, is true if and only if both P and Q are true. For example, "5 is prime and water boils at 212°F" is true because both parts are true.

The statement "P or Q," written $P \vee Q$, is true if one or both of the parts P and Q is true. For example, "The sun is made of gas or pigs fly" is true because the first part is true.

A negation symbol will only apply to the statement which immediately comes after it. For example, if P="Pigs fly" and W="Water boils at 212°F," then $\neg W \wedge P$="Water does not boil at 212°F and pigs fly." If we want to negate more, we use parentheses. For example, $\neg(W \wedge P)$="It is not true that both water boils at 212°F and pigs fly." The statements $\neg W \wedge P$ and $\neg(W \wedge P)$ are clearly different because the first is false and the second is true.

SOLVED PROBLEMS

Conjunctions

1. Suppose statements A, B, and F were true and C, D, and E were known false. Find the truth value (true or false) for each of the following:

   (a) $A \wedge C$

   (b) $(\neg B) \vee (\neg F)$

   (c) $A \wedge B$

   (d) $E \vee F$

   (e) $(A \wedge B) \vee C$

   (f) $\neg B \wedge D$

   (g) $\neg(B \wedge D)$

   (h) $(A \vee B) \wedge C$

   (i) $A \vee (B \wedge C)$

2. Suppose the statements "Jim drew a red card or Jim drew a seven" and "Jim bet $10 and Jim drew a red card" are both true. What can be said about the statements B="Jim bet $10," R="Jim drew a red card," and S="Jim drew a seven"?

Answers

1. (a) False
   (b) False
   (c) True
   (d) True
   (e) True
   (f) False
   (g) True
   (h) False
   (i) True

2. Because $B \wedge R$ = "Jim bet \$10 and Jim drew a red card" is true, we know that both B and R are true. We do not know if S is true or not; Jim could have drawn a red seven or else some other red card, and $R \vee S$ = "Jim drew a red card or Jim drew a seven" will still be true.

Truth Tables

A *truth table* lists out all the possible values (true or false) of several statements. For example, with two statements, A and B, there are four possibilities: both could be true, both could be false, A could be true and B false, or A could be false and B true. This is illustrated in Table 3-1, where T stands for true and F stands for false.

TABLE 3-1

A	B
T	T
T	F
F	T
F	F

This table can be extended to show the truth of compound statements $A \wedge B$, $A \vee B$, $\neg A$, and $\neg A \vee B$ under each of the four situations, as shown in Table 3-2. The third column, for instance, illustrates that $A \wedge B$ is true only when both A and B are true, and is false in all three of the other possibilities.

TABLE 3-2

A	B	$A \wedge B$	$A \vee B$	$\neg A$	$\neg A \vee B$
T	T	T	T	F	T
T	F	F	T	F	F
F	T	F	T	T	T
F	F	F	F	T	T

When two statements have the same true value under all possible situations, they are called *logically equivalent*, indicated by the symbol "≡." For example, Table 3-3 proves that $A \equiv \neg\neg A$.

TABLE 3-3

A	¬A	¬¬A
T	F	T
F	T	F

SOLVED PROBLEMS

Truth Tables

1. How many different truth possibilities are there with three statements A, B, and C? List them.
2. How many truth possibilities are there with (a) four statements, (b) five statements, (c) six statements, and (d) n statements?
3. Draw a truth table for two statements A and B. Make columns for $A \wedge B$, $\neg(A \wedge B)$, $\neg A$, $\neg B$, and $\neg A \vee \neg B$.
4. Identify equivalent statements from the table for problem (3).
5. Use a truth table to prove that $A \wedge (B \vee C) \equiv (A \wedge B) \vee (A \wedge C)$.
6. A *tautology* is a statement which is true in every case. Use a truth table to prove that $A \vee \neg A$ is a tautology.

Answers

1. There are four different possibilities for A and B with C true and four more with C false, for a total of eight. They can be listed as TTT, TTF, TFT, TFF, FTT, FTF, FFT, and FFF.
2. Each additional statement doubles the number of possibilities; thus, the answers are (a) 16, (b) 32, (c) 64, and (d) 2^n.
3. Shown in Table 3-4.
4. Table 3-4 proves that $\neg(A \wedge B) \equiv \neg A \vee \neg B$.

TABLE 3-4

A	B	$A \wedge B$	$\neg(A \wedge B)$	$\neg A$	$\neg B$	$\neg A \vee \neg B$
T	T	T	F	F	F	F
T	F	F	T	F	T	T
F	T	F	T	T	F	T
F	F	F	T	T	T	T

5. This requires a truth table with the eight different possibilities for A, B, and C, as shown in Table 3-5.

TABLE 3-5

A	B	C	$B \vee C$	$A \wedge (B \vee C)$	$A \wedge B$	$A \wedge C$	$(A \wedge B) \vee (A \wedge C)$
T	T	T	T	T	T	T	T
T	T	F	T	T	T	F	T
T	F	T	T	T	F	T	T
T	F	F	F	F	F	F	F
F	T	T	T	F	F	F	F
F	T	F	T	F	F	F	F
F	F	T	T	F	F	F	F
F	F	F	F	F	F	F	F

6.   Shown in Table 3-6.

TABLE 3-6

A	$\neg A$	$A \vee \neg A$
T	F	T
F	T	T

The Nature of Mathematical Proof

A *theorem* is a mathematical statement that can be proven true. A *proof* is a logical argument which begins with already-established statements and combines them according to certain rules of logic to obtain the desired statement. Because a proof requires already-established statements, there is no way to prove the very first statements of mathematics. Thus, the foundations of mathematics rest upon statements called *postulates* or *axioms*, which are accepted as true without proof. Attempts are made to make these postulates as basic and believable as possible. For example, two of the postulates used to prove statements about the natural numbers are "There is a number called 1" and "Every number has a number that comes after it." If you accept the postulates as true, then the results of mathematics follow from them. If you reject one or more postulates and replace them with different postulates, then it may turn out that a different sort of mathematics will result.

Conditional Statements

Most of the statements in mathematics take the form "If . . . , then . . ." and are called *conditional statements*. For example, because the results of mathematics depend entirely upon believing the fundamental postulates, the statement "$2+2=4$" really should be read as "If the fundamental postulates of mathematics are true, then $2+2=4$."

In the conditional statement "If P, then Q," the statement P is called the *hypothesis* and Q is the *conclusion*. The truth table for "If P, then Q," often written $P \Rightarrow Q$, is shown in Table 3-7.

TABLE 3-7

P	Q	$P \Rightarrow Q$
T	T	T
T	F	F
F	T	T
F	F	T

Notice that the only way the statement $P \Rightarrow Q$ could be false is if P is true but Q is false. For example, a friend who says, "If I have $1,000,000, then I'll buy you a car," is probably telling the truth. The only way for this to be false would be if your friend had $1,000,000 but did not buy you a car. Until your friend gets the money, you have no reason to believe that he or she was not telling the truth. Put another way, a conditional $P \Rightarrow Q$ is false only when P is true and Q is false. In particular, if P is false, then $P \Rightarrow Q$ is true no matter what Q is. Moreover, if Q is true, then $P \Rightarrow Q$ is true no matter what P is.

SOLVED PROBLEMS

Conditional Statements

1.   For each of the following conditional statements, state the hypothesis, the conclusion, and whether or not the conditional statement is true:

   (a)   If 7 is an even number, then 7 can be divided evenly by 2.
   (b)   If 7 is a prime number, then 9 is a prime number.
   (c)   If water boils at 212°F, then granite is a metamorphic rock.
   (d)   If 2 is a negative number, then 2^2 is a positive number.
   (e)   Water will freeze if the temperature goes below 32°F.

2.   Use a truth table to prove that the conditional statement $P \Rightarrow Q$ is logically equivalent to $\neg P \vee Q$.

   Answers

   1.   (a)   The hypothesis is "7 is an even number," and the conclusion is "7 can be divided by 2." This conditional statement is true because the hypothesis is false.
      (b)   The hypothesis is "7 is a prime number," and the conclusion is "9 is a prime number." This conditional statement is not true because the hypothesis is true and the conclusion is false.
      (c)   The hypothesis is "water boils at 212°F," and the conclusion is "granite is a metamorphic rock." Logically, this conditional statement is true because the two parts are both true. This statement does not say that granite is a metamorphic rock because water boils at 212°F, only that if the first part is true then the second part will also be true.
      (d)   The hypothesis is "2 is a negative number," and the conclusion is "2^2 is a positive number." This conditional statement is true because the hypothesis is false.
      (e)   This statement can be rewritten as follows: "If the temperature goes below 32°F, then water will freeze." Thus the hypothesis is "the temperature goes below 32°F," and the conclusion is "water will freeze." This conditional statement is true because the conclusion is true whenever the hypothesis is true.

   2.   Using the truth table in Table 3-8, we see that $P \Rightarrow Q \equiv \neg P \vee Q$ because the third and fifth columns have the same truth value in every possible situation.

   TABLE 3-8

P	Q	$P \Rightarrow Q$	$\neg P$	$\neg P \vee Q$
T	T	T	F	T
T	F	F	F	F
F	T	T	T	T
F	F	T	T	T

Contrapositives and Converses

The negation of a conditional statement $P \Rightarrow Q$ is not another conditional statement but the compound statement $P \wedge \neg Q$. For example, the negation of "If $x^2 = 9$, then $x = 3$" is "$x^2 = 9$ and $x \neq 3$."

   Every conditional statement $P \Rightarrow Q$ is logically equivalent to its *contrapositive* $\neg Q \Rightarrow \neg P$, as is shown in Table 3-9. For example, the contrapositive of "If the temperature drops below freezing, then this rain will turn to snow" is "If this rain does not turn to snow, then the temperature did not drop below freezing." Sometimes proving a statement's contrapositive is easier than proving the original statement directly.

   A common mistake is to confuse the contrapositive of a statement $P \Rightarrow Q$ with the *converse*: $Q \Rightarrow P$. For example, "If you are ten years old, then you cannot legally drive a car" is a true statement. The converse, "If you cannot legally drive a car, then you are ten years old," is false (it is not true in every instance).

TABLE 3-9

P	Q	$P \Rightarrow Q$	$\neg Q$	$\neg P$	$\neg Q \Rightarrow \neg P$
T	T	T	F	F	T
T	F	F	T	F	F
F	T	T	F	T	T
F	F	T	T	T	T

The contrapositive is "If you can legally drive a car, then you are not ten years old." This contrapositive is true no matter to whom the "you" refers, just as with the original statement.

SOLVED PROBLEMS

Converses and Contrapositives

1. For each of the conditional statements below, state the negation, contrapositive, and converse:

    (a) If there is no oxygen, then the candle will go out.
    (b) If water freezes, then the temperature is below 32°F.
    (c) If x^2 is even, then x is even.
    (d) If there is a Euler circuit, then every vertex has even degree.

Answers

1.   (a) Negation: "There is no oxygen, and the candle will not go out."
              Contrapositive: "If the candle will not go out, then there is oxygen."
              Converse: "If the candle will go out, then there is no oxygen."

      (b) Negation: "Water freezes, and the temperature is not below 32°F."
              Contrapositive: "If the temperature is not below 32°F, then water will not freeze."
              Converse: "If the temperature is below 32°F, then water freezes."

      (c) Negation: "x^2 is even, and x is not even."
              Contrapositive: "If x is not even, then x^2 is not even."
              Converse: "If x is even, then x^2 is even."

      (d) Negation: "There is a Euler circuit, and not every vertex has even degree."
              Contrapositive: "If not every vertex has even degree, then there is not a Euler circuit."
              Converse: "If every vertex has even degree, then there is a Euler circuit."

Comparing the World of Mathematics to Cartoons

In many ways, the world of mathematics is a simplified version of the real world, much like the world of cartoons and comic strips. For one thing, both deal with symbolic representations of idealized objects. It is impossible to draw a perfect circle, for example, yet people will immediately recognize what Fig. 3-1(a) represents. Similarly, pancakes in cartoons always come on plates, in tall stacks, with a square pat of butter, and with the syrup spilling over the edges, as depicted in Fig. 3-1(b). A real stack of pancakes like this would be difficult to eat; there are too many pancakes, the butter is not spread out, and there needs to be syrup on each individual pancake. However, it would thrill a person to be served a plate like this because it is immediately recognizable as the ideal stack of pancakes.

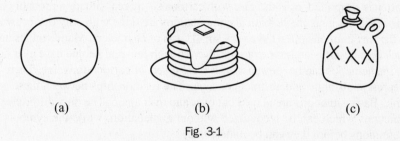

Fig. 3-1

In mathematics, everything is certain. For example, when 10 is divided by 2, the result is always 5. In real life, however, few things can be generalized with certainty. It cannot be said, for example, that everyone has ten toes. In cartoons, many things are certain. For example, anyone drinking the contents of the jug in Fig. 3-1(c) is guaranteed to get completely drunk. One syllable will then convey intoxication: "Hic!"

In mathematics, everything is either true or false. When numbers are plugged in for the variables of an equation, for example, they either make an equality or else do not. In comics, many things exist only in absolutes like this. For example, the cup in Fig. 3-2(a) contains hot, delicious coffee, while the cup in Fig. 3-2(b) is empty. A lukewarm cup of coffee is impossible in a comic strip. Similarly, beer in comics exists in only two states: full and frothy, as illustrated in Fig. 3-2(c), or else empty (usually on its side), as in Fig. 3-2(d).

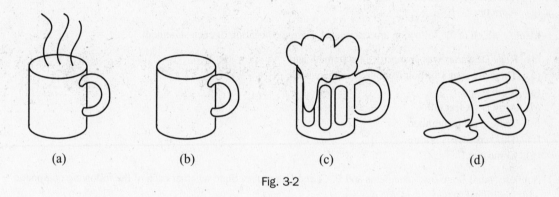

Fig. 3-2

In real life, we do not expect all statements to be either true or false. If a person says, "I am lying," we are more inclined to ignore the person than to be paralyzed by the paradox. However, when Kurt Gödel in 1930 wrote in mathematical language something roughly equivalent to "This theorem cannot be proven," he shook the very foundations of math and logic. Even though mathematicians still believe that every theorem is either true or false, Gödel's theorem revealed the possibility that there might never be a proof one way or the other.

In cartoons, dogs chase cats, cats chase mice, and mice are obsessed with cheese, which is always Swiss and always cut in wedges, as illustrated in Fig. 3-3. With the strategic placement of a cat, dog, mouse, and block of cheese, a single frame of a comic strip can illustrate a complicated sequence of events that is about to take place. Similarly, the objects in mathematics always behave in predictable ways: 2 always comes after 1, 7 is always 3 more than 4, and the multiplication in $5+2\times7$ is always performed before the addition.

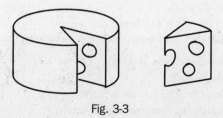

Fig. 3-3

Mathematics often is simplified, as is the case with cartoons. Any real situation probably has thousands of unknowns and variables, but most mathematical computations are done with only one or two variables. This is very similar to the way cartoons often have very small casts of characters. Many comics have featured one or two people stuck on a desert island, for example, even though few real people have ever been so isolated.

In conclusion, mathematics can be viewed as if it were a sort of cartoon universe. There are reoccurring symbols which represent absolute and abstract concepts. The relationships between these concepts are certain and predictable. Basic situations are played out over and over again. The only difference is that cartoons are drawn from pictures which can be recognized without explanation, while the symbols in mathematics require precise definitions before they can be understood.

SUPPLEMENTAL PROBLEMS

1. In 25 words or less, find a definition (answers will vary) for the following:

   (a) A *set*
   (b) The *domain* of a function
   (c) *Similar triangles*
   (d) *Parallel* lines
   (e) A *door*
   (f) A *bag*
   (g) A *spoon*

2. Identify which of the following are statements. Give the negation of each statement.

   (a) René Descartes was primarily a mathematician.
   (b) There are over 1 billion computers in the world.
   (c) Math is fun.
   (d) Amy is six feet tall.
   (e) Ten is a prime number.
   (f) This sentence is not true.
   (g) Come here!

3. Suppose A and D are true statements and B, C, and E are false. State whether each of the following compound and conditional statements is true or false:

   (a) $A \wedge B$
   (b) $A \vee B$
   (c) $A \Rightarrow B$
   (d) $A \wedge \neg C$
   (e) $\neg(B \wedge C)$
   (f) $A \vee (B \wedge C)$
   (g) $\neg(A \vee E) \Rightarrow (D \wedge \neg E)$

4. Use a truth table to prove the following:

   (a) $\neg(A \vee B) \equiv \neg A \wedge \neg B$
   (b) $A \vee (B \wedge C) \equiv (A \vee B) \wedge (A \vee C)$

5. Make a truth table for a conditional statement, $P \Rightarrow Q$, and its converse, $Q \Rightarrow P$. In what cases do the two statements have the same truth value?

6. Use a truth table to prove that $(A \wedge B) \Rightarrow A$ is a tautology.

7. For each conditional statement, state the hypothesis, conclusion, negation, contrapositive, and converse:

   (a) If the sun rises in the east, then the sun sets in the west.
   (b) If pigs fly, then dogs can speak English.
   (c) If rocks are flexible, then George Washington was a president.
   (d) If trees are made from carbon dioxide, then trees are not heavy.

(e) If the angles of a triangle sum to 180°, then the area of a circle is πr^2.

(f) There will be no alcohol if the party is at the church.

(g) If x is an integer, then x is a rational number.

Answers

1. (a) A *set* is a collection of objects.
   (b) The *domain* of a function is the set of all things to which the function relates elements of the range.
   (c) When two triangles have the same three angles, they are called *similar triangles*.
   (d) Lines in the same plane are *parallel* if they can be extended forever and never intersect.
   (e) A *door* is a large flat object which can be moved to block an entrance in a wall.
   (f) A *bag* is a small flexible container used for storing and carrying objects.
   (g) A *spoon* is a utensil that ends in a round, concave oval and is used for stirring and scooping soft foods and liquids.

2. (a) This is a statement, with negation "René Descartes was not primarily a mathematician."
   (b) This is a statement, with negation "There are not over 1 billion computers in the world."
   (c) This is an opinion, not a statement.
   (d) This is not a statement unless it is clear which Amy is meant, in which case the negation will be "Amy is not six feet tall."
   (e) This is a statement, with negation "Ten is not a prime number."
   (f) This is not a statement because it cannot be either true or false.
   (g) This is a command, not a statement.

3. (a) False
   (b) True
   (c) False
   (d) True
   (e) True
   (f) True
   (g) True

4. (a) $\neg(A \vee B) \equiv \neg A \wedge \neg B$ is true because the fourth and seventh columns of the truth table (Table 3-10) have the same truth value in each case.

TABLE 3-10

A	B	$A \vee B$	$\neg(A \vee B)$	$\neg A$	$\neg B$	$\neg A \wedge \neg B$
T	T	T	F	F	F	F
T	F	T	F	F	T	F
F	T	T	F	T	F	F
F	F	F	T	T	T	T

(b) $A \vee (B \wedge C) \equiv (A \vee B) \wedge (A \vee C)$ is true because the fifth and eighth columns of the truth table (Table 3-11) have the same truth value in every case.

TABLE 3-11

A	B	C	$B \wedge C$	$A \vee (B \wedge C)$	$A \vee B$	$A \vee C$	$(A \vee B) \wedge (A \vee C)$
T	T	T	T	T	T	T	T
T	T	F	F	T	T	T	T
T	F	T	F	T	T	T	T
T	F	F	F	T	T	T	T
F	T	T	T	T	T	T	T
F	T	F	F	F	T	F	F
F	F	T	F	F	F	T	F
F	F	F	F	F	F	F	F

5. The truth table (Table 3-12) shows that $P \Rightarrow Q$ and its converse $Q \Rightarrow P$ are both true only when P and Q have the same truth value (either true or false). There is no situation for which a conditional statement and its converse are both false.

TABLE 3-12

P	Q	$P \Rightarrow Q$	$Q \Rightarrow P$
T	T	T	T
T	F	F	T
F	T	T	F
F	F	T	T

6. The truth table (Table 3-13) shows that $(A \wedge B) \Rightarrow A$ is true in every case; thus, it is a tautology.

TABLE 3-13

A	B	$A \wedge B$	$(A \wedge B) \Rightarrow A$
T	T	T	T
T	F	F	T
F	T	F	T
F	F	F	T

7. (a) Hypothesis: "The sun rises in the east."
       Conclusion: "The sun sets in the west."
       Negation: "The sun rises in the east, and the sun does not set in the west."
       Contrapositive: "If the sun does not set in the west, then the sun does not rise in the east."
       Converse: "If the sun sets in the west, then the sun rises in the east."

    (b) Hypothesis: "Pigs fly."
       Conclusion: "Dogs can speak English."
       Negation: "Pigs fly, and dogs cannot speak English."
       Contrapositive: "If dogs cannot speak English, then pigs cannot fly."
       Converse: "If dogs speak English, then pigs fly."

    (c) Hypothesis: "Rocks are flexible."
       Conclusion: "George Washington was a president."
       Negation: "Rocks are flexible, and George Washington was not a president."
       Contrapositive: "If George Washington was not a president, then rocks are not flexible."
       Converse: "If George Washington was a president, then rocks are flexible."

(d) Hypothesis: "Trees are made from carbon dioxide."
Conclusion: "Trees are not heavy."
Negation: "Trees are made from carbon dioxide, and trees are heavy."
Contrapositive: "If trees are heavy, then trees are not made from carbon dioxide."
Converse: "If trees are not heavy, then trees are made from carbon dioxide."

(e) Hypothesis: "The angles of a triangle sum to $180°$."
Conclusion: "The area of a circle is πr^2."
Negation: "The angles of a triangle sum to $180°$, and the area of a circle is not πr^2."
Contrapositive: "If the area of a circle is not πr^2, then the angles of a triangle do not sum to $180°$."
Converse: "If the area of a circle is πr^2, then the angles of a triangle sum to $180°$."

(f) Hypothesis: "The party is at the church."
Conclusion: "There will be no alcohol."
Negation: "The party is at the church, and there will be alcohol."
Contrapositive: "If there is alcohol, then the party is not at the church."
Converse: "If there is not alcohol, then the party is at the church."

(g) Hypothesis: "x is an integer."
Conclusion: "x is a rational number."
Negation: "x is an integer, and x is not a rational number."
Contrapositive: "If x is not a rational number, then x is not an integer."
Converse: "If x is a rational number, then x is an integer."

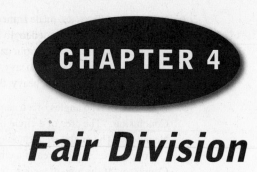

Fair Division

Fair division is the challenge of sharing something among several people in such a way that each person finds his or her share reasonable. This can be easy if the people need to share money, or a large number of identical objects (for example, a bucket of nails). Things are also easy if everyone chooses to let an outside authority make the decision. For example, a group of children might have a parent divide something among them. In this chapter, we will suppose that the people have to decide among themselves how to share something that does not have an obvious way of being divided.

One way to approach fair division is to view the process of sharing things as a game. The people are called *players* and take *turns* dividing the *winnings* (the items to be shared) based on *rules* that everyone agrees to follow.

Sharing among Two People

When there are only two players, there is a game with simple rules:

1. Flip a coin to decide who will be player 1 and who will be player 2.
2. Player 1 divides the winnings into two shares that seem fair to her.
3. Player 2 takes whichever share seems best to him.
4. Player 1 gets the remaining share.

For example, two children can share a cookie if one of them breaks it into two pieces and the other one chooses first. Even if both kids wanted the whole cookie, they might feel better if they view the sharing as a game and the resulting half-cookies as their winnings.

Because the job of dividing something into two equal shares is more difficult than the job of evaluating which share is greater, player 2 nearly always has an advantage. It is for this reason that the game begins by naming players 1 and 2 by a coin flip.

In ancient Greek mythology, Prometheus and Zeus played this game to determine how animal sacrifices should be conducted. Prometheus divided the animal into two piles—one was a pile of scraps of meat, and the other consisted of all the bones and innards wrapped in a layer of fat to look like a juicy package. Zeus was tricked into choosing the latter pile, and thus ever afterward the Greeks ate the meat from their sacrifices and burned the rest as an offering to the gods.

There are two difficulties with this game. First, it only works with two players. Second, it only works with things that are still desirable when split in half (cakes, for example, and not cats).

SOLVED PROBLEMS

Sharing among Two People

1. Describe how two children could use this method to divide a pile of blocks.
2. How could a married couple use this method to share chores?

3. Could two sisters who lived in the same home use this method to share a car?

4. What could go wrong if three kids tried to divide a pile of toy cars in the following way: player 1 divides the cars into three piles, player 2 picks one pile, player 3 picks another pile, and player 1 takes the last pile?

Answers

1. One child divides the blocks into two piles, and the other chooses a pile with which to play.

2. One spouse writes down all the chores that need to be done into two columns. The other spouse then does all the chores from one column. It would probably be a good idea to repeat this game frequently, and not force each person to always do the same chores.

3. While it is not possible to cut a car into two usable pieces, it might be possible to divide each week or month. One sister could take a daily planner and highlight the hours into two categories. The other sister would then choose the collection of times she liked better.

4. Players 1 and 2 could team up to cheat player 3. Player 1 could put one car each into the first two shares and all the rest into the third pile. Player 2 would then take the big pile. Player 3 would only get a single car. Player 1 would only get a single car, too. However, after player 3 runs off crying, players 1 and 2 can put their piles together and play with almost all the cars.

Sharing among Three People

Here is a method for sharing with three people:

1. Randomly label the players 1, 2, and 3 (pick numbers from a hat).
2. Player 1 divides the winnings into three piles.
3. Player 1 then asks players 2 and 3, "Does this look fair to you?"

   (a) If players 2 and 3 both say yes, player 2 picks first and then player 3.
   (b) If only player 2 says, "No," player 2 picks first, and then player 3.
   (c) If only player 3 says, "No," player 3 picks first, and then player 2.
   (d) If both of them say, "No," then they each pick a share, put them together into one pile, and divide it using the two-person sharing game.

4. Player 1 takes whatever pile is left.

The strategy for this game is easy. If you are player 1, you will have to make three piles that all look fair to you. You are guaranteed to get one of them. If you are player 2 or 3, count the number of piles that look fair (at least a third of the total). If there are two reasonable piles, say, "Yes, it looks fair," and you will get one of them. If only one pile seems fair, say, "Not fair!" You will either get this whole pile to yourself or else get to share more than two-thirds of the winnings with one other player.

SOLVED PROBLEMS

Sharing among Three People

1. Suppose you are player 3 and player 1 has divided a round pie in one of the ways illustrated in Fig. 4-1. How can you ensure that you get at least a third of the pie?

   ### Answers

   1. (a) Say that the division is fair. Player 2 will pick first, but you will get one of these three equal pieces. This division, of course, is the only one which guarantees a fair slice for player 1. The other four situations reflect where player 1 either does not care for pie or is trying to cheat you.

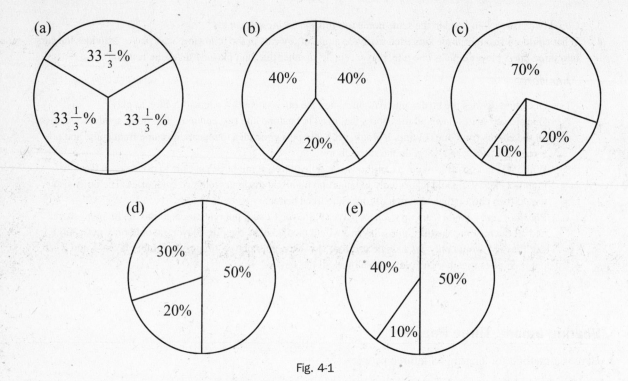

Fig. 4-1

(b) Say that the division is fair. Whether player 2 complains or not, you will be able to take a 40% slice, which is more than a fair third. If you complain, then player 2 might also complain. In this case, you and player 2 would each pick a slice, put them together, and then split them by the two-player game. Player 2 could cheat you by picking the 20% slice. No matter which slice you took, the total would only be 60%, so player 2 could limit you to only 30%, not a fair share. This is avoided if you declare the split to be fair.

(c) Because there is only one slice over a third, you should declare this unfair. If player 2 thinks it is fair, you will get to choose first, and can take the 70% slice. If player 2 says it is unfair, then make sure to pick the 70% slice to share. Even if player 2 picks the 10% slice, you will still end up with at least 40% of the pie.

(d) With only one slice above $33\frac{1}{3}\%$, declare this unfair. If you get to pick first (player 2 declares this fair), take the 50% slice. If you have to share, the combined piece will be either 80% or 70%, depending on what player 2 picks (assuming you take the 50% slice). Thus, you will end up with at least either 40% or 35%.

(e) Here, you should declare the cut fair. Player 2 will pick first and probably take the 50% slice, but you will get at least 40%, which is more than a fair third. This division is not called *envy free* because even though you get more than your fair share, someone else gets even more than you.

The Last-Diminisher Method

Here is a method that works for any number of players sharing something that is easily divisible:

1. Randomly assign an order to all the players.
2. Player 1 takes a single share out of the whole pile.
3. Each other player, in order, decides either of the following:

   (a) That the share is reasonable, and leaves it alone
   (b) That the share is too big, and cuts it down to a smaller share

4. When all players have had their say, the share goes to the last player who reduced it.
5. The players who have not yet received a share begin the game over again with what remains of the winnings.

   The advantage of this game is that it will work with more than three players. The main disadvantage is that there can be a great amount of cutting, which can reduce a cake or pie to mush. Also, it can be very difficult to estimate a fair share early in the game. For example, if seven people want to share a cake, it can be very hard to look at a single slice and determine if it is more or less than a fair seventh.

   It sometimes happens that there is something that makes one share more valuable than any other. For example, children might fight over the tiniest piece of cake, so long as it contains the only frosting rose. In such a case, it is a good idea to put the rose aside and first divide the cake without it. After everyone has a fair share of cake, a new game can be played to share the rose.

SOLVED PROBLEMS

The Last-Diminisher Method

1. Explain how four kids might share a pile of toy cars by the last-diminisher method.
2. How could two people split a pineapple by the last-diminisher method?
3. Suppose ten highly argumentative people were to try to split a cake by the last-diminisher method. Suppose further that each share was made with a single cut of a knife, and each diminishing was again made by a single cut. If no one ever thought a slice was fair, how many cuts would be made before the entire cake was divided?
4. After World War II, the Allied powers wanted to divide control of Germany between the capitalists and the communists. However, both sides wanted any piece that contained Berlin. How was this arranged?

Answers

1. One kid, randomly chosen, separates out some of the cars for himself. If everyone else agrees this is a reasonable share, then he takes them. If anyone feels that he has taken too much, she may put some of the cars back into the main pile and take the remaining share for herself. Anyone who feels that her share is still too big may take the pile from her by putting back even more cars. When that share is settled, the three children who have not received a share then divide the remaining cars among themselves using the same process.
2. Player 1 cuts off part of the pineapple. If player 2 is willing to let player 1 have this, player 2 takes the rest of the pineapple and they are done. If player 2 thinks player 1 has taken too much, player 2 trims off some part of player 1's slice and takes the diminished serving. The trimmings, plus the rest of the pineapple, go to player 1.
3. Player 1 chooses a piece of the cake with one cut of the knife. Player 2 cuts off some, player 3 cuts off more, and so on. Player 10 trims with a tenth cut, and takes the resulting piece. The remaining cake, with all the shaved-off slivers, is squished back together for the other nine players to share. In the end, there will be $10+9+8+7+6+5+4+3+2=54$ cuts made to the cake. Note that when there is only one person left, there is no need to make a last cut.
4. Because Berlin was so contentious, the Allies divided everything else first. A line was drawn, separating East and West Germany. Even though Berlin was in the eastern half of the country, it was further divided into East Berlin for the communists and West Berlin for the capitalists.

Sharing the Indivisible by Making Bids

If people want to divide something that cannot be cut into pieces, for example a house, then money is needed for things to be fair. Either the house must be sold and the money divided equally, or else one person gets the house and pays everyone else a fair share of the house's value. The best way to calculate the value of a house is to sell it, but there are also realtors who can assess the market price.

Other items can be much harder to value. For example, how could two sisters divide their great-grandmother's cookbook? Sold to a used bookstore, it might be worth only $2. Neither one would give it up for the $1 half-share of fair market value. Instead, we let the players name their own values in a sort of silent auction called the *method of sealed bids*:

1. Each player writes down a value for each item, the highest amount he or she would pay for that object.
2. The value to each player of all the goods is totaled, and a fair share is calculated.
3. Each item goes to the player who values it highest (choose randomly in a tie).
4. Players who receive more in goods (as they value them) than their fair share must pay the difference in cash.
5. Players who receive less than their share receive the difference in cash.
6. Any cash leftover is split evenly among all players.

For example, suppose three brothers want to split their grandfather's pocket watch and cane among themselves. Andy values the watch at $300 and the cane at $30. David values the watch at $200 and the cane at $70. Samuel values the watch at $200 and the cane at $40.

Andy feels the total value of the goods is $330, so his fair share is a third: $110. He gets the watch because he bid highest for it. Because the watch is worth $300 to him, he must pay the $190 difference in cash.

David gets the cane, which is worth $70 to him. He values the goods at a total of $270, so his fair share is $90. The remaining $20 is paid to him from the cash that Andy paid out.

Samuel does not get either of the items. Because he values the goods at a total of $240, he gets his $80 one-third in cash.

At this point, each brother has received his fair share in some combination of money and/or goods. However, there is $90 left from the $190 that Andy put down after David took $20 and Samuel took $80. This is divided evenly: $30 each. The end result is that each brother receives more than he felt was his fair share. Table 4-1 illustrates the complete computation.

TABLE 4-1

	ANDY	DAVID	SAMUEL
Bid for watch	$300	$200	$200
Bid for cane	$30	$70	$40
Total cash value of bids	$330	$270	$240
Cash value of fair share	$110	$90	$80
Share	Watch ($300) − $190	Cane ($70) + $20	$80
Extra $\frac{1}{3}(190 - 80 - 20)$	$30	$30	$30
Final share	Watch ($300) − $160	Cane ($70) + $50	$110

In short, David gets the cane, Andy gets the watch, and Andy pays $50 to David and $110 to Samuel. If Andy had known that his brothers would only value the watch at $200, he could have lowered his bid to $201. This way, he would get the watch but pay out less in cash. It is to avoid manipulations like this that everyone must write down their bids in secret before computing each share.

SOLVED PROBLEMS

The Method of Sealed Bids

1. Suppose three pirates want to divide a ship, a treasure map, a gold crown, and a chest of 1,200 silver pieces among the three of them. If they bid on the four items as listed in Table 4-1, how will the method of sealed bids distribute the booty?

TABLE 4-2

	REDBEARD	BLACKBEARD	PETE
Ship	900 silver	600 silver	500 silver
Map	600 silver	500 silver	400 silver
Crown	300 silver	1,000 silver	900 silver
Chest	1,200 silver	1,200 silver	1,200 silver

2. Suppose two sisters, Maria and Isabelle, both want to inherit their great-grandmother's cookbook. Using the method of sealed bids, Maria values the book at $250 and Isabelle values it at $150. What is the outcome?

3. Four housemates decide to share the apartment chores using the method of sealed bids. They write down for each chore the weekly amount of money for which they would either do the chore or pay someone else to do it for them. The result is given in Table 4-3. If each chore is given to the person willing to do it for the least amount of money, how should the work be distributed?

TABLE 4-3

	ANNA	BOB	CARLOS	KIM
Wash kitchen	20	30	25	30
Wash bathroom	30	60	45	40
Take out trash	40	20	15	30
Sweep floors	30	10	15	20

Answers

1. Redbeard gets both the ship and the map. Blackbeard gets the crown. The 1,200 silver pieces are added to the shared cash at the end of the process. The table of computations is given in Table 4-4.

TABLE 4-4

	REDBEARD	BLACKBEARD	PETE
Ship	900 silver	600 silver	500 silver
Map	600 silver	500 silver	400 silver
Crown	300 silver	1,000 silver	900 silver
Chest	1,200 silver	1,200 silver	1,200 silver
Total value	3,000 silver	3,300 silver	3,000 silver
Value of fair share	1,000 silver	1,100 silver	1,000 silver
Share	Ship (900) + map (600) −500 silver	Crown (1,000) +100 silver	1,000 silver
Extra silver	200 silver	200 silver	200 silver
$\frac{1}{3}\left(1,200 + 500 - 100 - 1,000\right)$			
Final share	Ship + map − 300 silver	Crown + 300 silver	1,200 silver

To make up for the fact that Redbeard gets 1,500 silver worth of goods, he must add 500 silver to the treasure chest to make up his fair share. After Blackbeard takes his 100 and Pete his 1,000, there will still be $1,200 + 500 - 100 - 1,000 = 600$ silver left over. Thus, 200 pieces of bonus silver are added to each pirate's share.

2. The table of computations is illustrated in Table 4-5.

TABLE 4-5

	MARIA	ISABELLE
Cookbook	$250	$150
Fair share	$125	$75
Share	Book ($250) − $125	$75
Extra $\frac{1}{2}(125-75)$	$25	$25
Final share	Book − $100	$100

Here, Maria pays the $125 difference between the $250 cookbook she gets and the $125 that is her fair share. Isabelle takes $75 of this. The remaining $50 is split evenly between the two of them as bonus cash.

3. Table 4-6 contains all the computations.

TABLE 4-6

	ANNA	BOB	CARLOS	KIM
Wash kitchen	$20	$30	$25	$30
Wash bathroom	$30	$60	$45	$40
Take out trash	$40	$20	$15	$30
Sweep floors	$30	$10	$15	$20
Total value	$120	$120	$100	$120
Fair share	$30	$30	$25	$30
Share of chores	Washes kitchen ($20) and bathroom ($30), gets $20	Sweeps floors ($10), pays $20	Takes out trash ($15), pays $10	Pays $30
Extra cash $\frac{1}{4}(-20+20+10+30)$	$10	$10	$10	$10
Final chores	Washes kitchen and bathroom, gets $30	Washes floors, pays $10	Takes out trash	Pays $20

Here, the fair share is the dollar amount that each person feels would be a fair share of the chores. Thus, when Anna does $50 worth of chores, she deserves $20 in cash because she is doing more than her share. The other three pay out a total of $60, so the leftover $40 is distributed equally.

SUPPLEMENTAL PROBLEMS

1. Together, two friends win a prize of $100. How can they share it?
2. There is only one slice of coffee cake left. How can two roommates share it?
3. Mike and Kim have only one television set and want to watch different shows this evening. How can they share?
4. How can three sisters split a baseball card collection?
5. Suppose you are sharing a calzone with two other people. One of them cuts the calzone into two big slices and one small one. She then asks, "Is this fair?" What should you say?

6. Two housemates are moving out and want to divide all the furniture and decorations they collected together. Both of them value a certain painting higher than everything else combined. What can they do?

7. How can the last-diminisher method be used to share a pile of Halloween candy among four people?

8. Suppose five people want to lay out towels on a spot of crowded beach. How could the last-diminisher method be used to share the area?

9. Two people want to split a cat. Amy values the cat at $200. Karen values the cat at $300. How can they both be made happy?

10. Pedro, Jose, and Alex want to share an autographed baseball bat and glove. Using the method of sealed bids, they value the items as detailed in Table 4-7. How should the items be distributed?

TABLE 4-7

	PEDRO	JOSE	ALEX
Bat	$500	$800	$550
Glove	$475	$400	$500

11. Bruce, Michael, Wendy, and Scott have collectively inherited $10,000, a painting, and an old car. Using sealed bids, they value the items as detailed in Table 4-8. How should the inheritance be split?

TABLE 4-8

	BRUCE	MICHAEL	WENDY	SCOTT
$10,000	$10,000	$10,000	$10,000	$10,000
Painting	$2,000	$1,000	$3,000	$2,000
Car	$800	$1,000	$800	$1,200

12. Suppose Joe is in the process of using the method of sealed bids to share an antique clock and set of silver with his two sisters, Judy and Freda. When they are looking away, he sees their bids, detailed in Table 4-9.

   (a) How can he win the clock at minimum cost?
   (b) If he does not want either item, how can he maximize the cash he gets?

TABLE 4-9

	JUDY	FREDA
Clock	$1,100	$800
Silver	$1,000	$1,600

13. Suppose housemates Sarah, Alyssa, and Joey want to share the household chores using the method of sealed bids. They set the monthly value of each chore as given in Table 4-10. How should the chores be shared?

TABLE 4-10

	SARAH	ALYSSA	JOEY
Dishes	$50	$60	$40
Laundry	$30	$40	$40
Floors	$20	$20	$10
Bathroom	$20	$30	$30

Answers

1. Money is easily divided: each friend gets $50.

2. They should flip a coin to determine who should cut the cake. The other person gets to take either one of the two pieces.

3. They should divide the half-hour time blocks of the evening into two sets of times. Mike will choose the channel during one set of times, and Kim during the other. To divide the evening, one of them (chosen by coin flip) should write up the two sets of times, and the other one should pick one of those sets.

4. If the sisters use the three-person method, they should randomly label themselves as player 1, 2, and 3. Player 1 divides the cards into three piles, then asks, "Is this fair?" If one of the others says "No," then she picks first, followed by the other. If both sisters say, "No," then they each pick a pile to put together and divide by the two-person method. If both sisters say, "Yes," then player 2 takes a pile, and then player 3. In any case, player 1 gets the last pile remaining.

   If the sisters use the last-diminisher method, they also arrange themselves into players 1, 2, and 3. Player 1 takes out a share of the cards. If player 2 wants some of those cards, she can remove some cards from the pile and take the remaining ones as her share. Otherwise, player 2 leaves the share with player 1. If player 3 wants some of the share, she can take even fewer as her share. In any case, the share goes to whomever diminished the pile last. The two sisters who have not yet selected shares then repeat the process between themselves with the remaining baseball cards.

5. You should say, "Yes, it is fair," because she will pick last, and thus will probably end up with the small piece. You will be able to pick one of the big ones.

6. The housemates should put the painting aside and divide the rest of the items using the one-person-divides-and-the-other-picks-first method. When that is done, they can figure out what to do with the painting. If they are willing to send the painting back and forth, they could divide the months of the year into two sets, during which each person could display the painting at his or her house. If this is too difficult, they can value the painting using the method of sealed bids: the person who gets the painting will pay for the other person's share with cash.

7. One person (chosen randomly) lays out a share of the candy. Any of the other three people can diminish the pile by taking some candy out. The share goes to whomever does this last. When that share is settled, the three people without candy then repeat the process with the remaining Halloween candy.

8. The five people are randomly assigned numbers 1 through 5. Player 1 then takes a stick and outlines an area on the beach that he or she feels would be a reasonable spot for a towel. The other four players, in order, decide to either leave the spot alone or else make the spot smaller (draw a new area inside the old one). The last person to diminish the spot gets to set up his or her towel in that area. The other four people then repeat the process to divide the remaining beach.

9. Karen should get the cat because she values it more. She feels that her fair share of the $300 cat should be $150, so she should pay back the extra $150 with cash. Amy feels that her fair share of the $200 cat is $100, so she takes $100 of this money. The remaining $50 is split evenly between them. Thus, Karen gets the cat and pays Amy $125. Because both of them receive $25 more than what they calculated as a fair share, they should be happy.

10. The bat goes to Jose, and the glove to Alex. The computations are given in Table 4-11.

TABLE 4-11

	PEDRO	JOSE	ALEX
Bat	$500	$800	$550
Glove	$475	$400	$500
Total	$975	$1,200	$1,050
Share	$325	$400	$350
Receive	$325	Bat ($800) − $400	Glove ($500) − $150
Extra $\frac{1}{3}\left(400 + 150 - 325\right)$	$75	$75	$75
Final share	$400	Bat − $325	Glove − $75

11. Here, Wendy gets the painting and Scott gets the car. The extra money is calculated by taking the $10,000 and subtracting the $3,200+$3,000+$450+$2,100 required to meet everyone's idea of a fair share. This leftover $1,250 is split four ways, giving each person an extra $312.50. The details are given in Table 4-12.

TABLE 4-12

	BRUCE	MICHAEL	WENDY	SCOTT
$10,000	$10,000	$10,000	$10,000	$10,000
Painting	$2,000	$1,000	$3,000	$2,000
Car	$800	$1,000	$800	$1,200
Total	$12,800	$12,000	$13,800	$13,200
Fair share	$3,200	$3,000	$3,450	$3,300
Receive	$3,200	$3,000	Painting ($3,000)+$450	Car ($1,200)+$2,100
Extra money	$312.50	$312.50	$312.50	$312.50
Final shares	$3,512.50	$3,312.50	Painting+$762.50	Car+$2,412.50

12. (a) To win the clock at minimum cost, Joe wants to make a bid just slightly higher than anyone else's, for example $1,101. In order to make his total share as large as possible, he then wants to bid as high as possible on the silver without winning it, for example $1,599. The results are then calculated in Table 4-13.

TABLE 4-13

	JUDY	FREDA	JOE
Clock	$1,100	$800	$1,101
Silver	$1,000	$1,600	$1,599
Total	$2,100	$2,400	$2,700
Share	$700	$800	$900
Receive	$700	Silver ($1,600)−$800	Clock ($1,101)−$201
Extra $\frac{1}{3}(800+201-700)$	$100.33	$100.33	$100.33
Final share	$800.33	Silver−$699.67	Clock−$100.67

Joe wins the clock and only has to pay out $100.67.

(b) To maximize his cash, Joe should bid as high as possible for each item without winning the bid. This is computed in Table 4-14.

TABLE 4-14

	JUDY	FREDA	JOE
Clock	$1,100	$800	$1,099
Silver	$1,000	$1,600	$1,599
Total	$2,100	$2,400	$2,698
Share	$700	$800	$899.33
Receive	Clock ($1,100)−$400	Silver ($1,600)−$800	$899.33
Extra $\frac{1}{3}(400+800-899.33)$	$100.22	$100.22	$100.22
Final share	Clock−$299.78	Silver−$699.78	$999.55

Notice that in both instances, there will be a penny left over after the final shares are given out. Hopefully no one will fight over it. However, Judy and Freda ought to be made suspicious by how close Joe's bids were to their own.

13. The chores are assigned to the people willing to do them for the least amount of money. The calculations and end results are given in Table 4-15.

TABLE 4-15

	SARAH	ALYSSA	JOEY
Dishes	$50	$60	$40
Laundry	$30	$40	$40
Floors	$20	$20	$10
Bathroom	$20	$30	$30
Total	$120	$150	$120
Share	$40	$50	$40
Chores	Do laundry ($30), clean bathroom ($20), receive $10	Pay $50	Do dishes ($40), clean floors ($10), receive $10
Extra money $\frac{1}{3}(50-10-10)$	$10	$10	$10
Final chores	Do laundry and bathroom, receive $20	Pay $40	Do dishes and floors, receive $20

CHAPTER 5

Functions

A *function* is a way to relate the elements of two sets. The first set is called the *domain*, and the second set the *range*. The only requirement is that the function must take each element of the domain to exactly one element of the range.

Computing with Functions

Most of the functions in our lives are generally invisible. For example, when you type people's names into a database to look up their information, the computer uses a function to relate the list of all people's names (the domain) with the list of all personal information (the range).

Many of the buttons on a calculator are functions. The x^2 button, for example, represents a function. If you type "5" and then hit x^2, the calculator outputs "25," the number related to 5 under the *square number* function.

To represent a function mathematically, we need to know three things: (1) what will the function be called (the name), (2) what is going to be plugged into the function (the variable), and (3) what will the function do with the variable (the formula)?

The name of the function is written first. Because mathematicians are lazy, most functions have one-letter names. The most common are *f*, *g*, and *h*. The trigonometric functions have three-letter names, for example *sin*, *cos*, and *tan*.

The variable which represents the number being plugged into the function is put in parentheses immediately after the function name. The variable is usually *x*, for example $f(x)$. Because *f* is a name and not a number, this does not represent multiplication: $f(x)$ is pronounced "*f* of *x*" and not "*f* times *x*."

After the name and variable is an equals sign, followed by a detailed explanation of what the function will do with the variable. The square number function is thus $f(x) = x^2$.

To evaluate a function, just replace each variable with the number in parentheses. For example, if $g(x) = 3x^2 - 4x + 2$, then $g(5) = 3(5)^2 - 4(5) + 2 = 57$ and $g(-2) = 3(-2)^2 - 4(-2) + 2 = 22$.

Similarly, if $h(t) = \dfrac{t^2 - 5\sqrt{t}}{t - 3}$, then $h\left(2 + \sqrt{3}\right) = \dfrac{\left(2 + \sqrt{3}\right)^2 - 5\sqrt{(2 + \sqrt{3})}}{\left(2 + \sqrt{3}\right) - 3}$. This may look complicated,

but the only thing that happened was that each variable *t* was replaced by the expression $\left(2 + \sqrt{3}\right)$.

SOLVED PROBLEMS

Computing with Functions

Suppose $f(x) = 3x - \sqrt{x}$, $g(s) = \dfrac{1}{s}$, and $Bob(t) = \dfrac{t^3 - 1}{5 - t}$. Use these functions to evaluate the following:

1. $f(4)$
2. $f(5)$
3. $g(-2)$
4. $Bob(3)$
5. $f\left(2 - \sqrt{3}\right)$ (do not simplify)
6. $g(W)$
7. $Bob(Fred)$
8. $g(f(x))$

Answers

1. $f(4) = 3(4) - \sqrt{4} = 12 - 2 = 10$

2. $f(5) = 3(5) - \sqrt{5} = 15 - \sqrt{5}$

3. $g(-2) = \dfrac{1}{-2} = -\dfrac{1}{2}$

4. $Bob(3) = \dfrac{3^3 - 1}{5 - 3} = \dfrac{26}{2} = 13$

5. $f(2 - \sqrt{3}) = 3\left(2 - \sqrt{3}\right) - \sqrt{(2 - \sqrt{3})}$

6. No matter what W may be, it replaces each s in the formula for $g(s)$; thus, $g(W) = \dfrac{1}{W}$.

7. $Bob(Fred) = \dfrac{\left(Fred\right)^3 - 1}{5 - \left(Fred\right)}$. This cannot be simplified unless we know

   what *Fred* represents.

8. $g(f(x)) = \dfrac{1}{f(x)}$. Here we know what $f(x)$ equals; thus, we can write the following:

   $g(f(x)) = \dfrac{1}{f(x)} = \dfrac{1}{\left(3x - \sqrt{x}\right)}$. It is called *composition* when one function is plugged into another like this.

   Often, $g(f(x))$ is written $(g \circ f)(x)$.

Graphing Functions

In the early 1600s, French mathematician and philosopher René Descartes had a realization while lying in bed and watching flies crawl about on the tiles of the ceiling: the position of a fly could be described by counting tiles horizontally and vertically. For example, the fly in Fig. 5-1(a) is on tile $(2, 4)$, the one that is two tiles to the right and four up from the lower left corner.

This realization formed the basis of the *Cartesian plane*, a plane where every point is associated with a pair of numbers: a horizontal *x-coordinate* and a vertical *y-coordinate*. For example, in Fig. 5-1(b), the points $(2, 4)$ and $(3.5, 1)$ are illustrated.

For functions, it is natural to let the x-coordinate be the number plugged into the function and the y-coordinate be the result that comes out of the function. If the function is called f, then this is written $y = f(x)$. For example, if all of the infinitely many points between -1 and 3 are plugged into the function

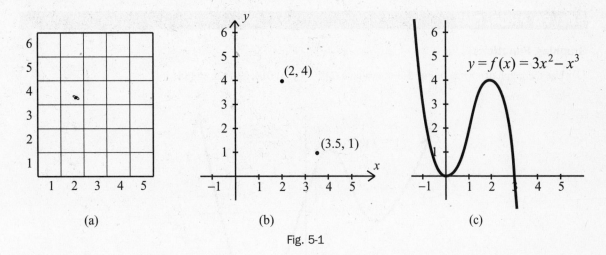

Fig. 5-1

$f(x) = 3x^2 - x^3$ and plotted on the Cartesian plane, they will form the continuous curve illustrated in Fig. 5-1(c). This is the *graph* of the function.

There are many tricks to graphing. One is to plot several points and then connect them with a curve. Another is to have a graphing calculator plot hundreds of points within a specified domain. Algebra can be used to find where the function crosses the x-axis (where $y = 0$). Calculus can be used to find the high and low points on the graph. In any case, it helps to anticipate the shape of the graph.

A *constant* function is one where $f(x) = a$ for some constant a, for example $f(x) = 3$ or $K(t) = 4$. The graph of a constant function will always be a horizontal line, as illustrated in Fig. 5-2(a).

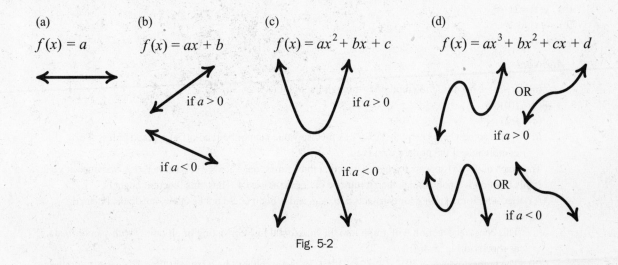

Fig. 5-2

A *linear* function is one where the highest power of the variable is 1, for example $g(x) = 3x + 2$ or $h(x) = -8x + 3$. The graph will be a straight line, as illustrated in Fig. 5-2(b). To find the graph, it is enough to plot two points, and then draw the straight line through them.

A *quadratic* function is one where the highest power of the variable is 2, for example $s(t) = -16t^2 - 20t + 112$ or $F(x) = x^2 - 3x + 4$. The shape of the graph will be a *parabola*, as illustrated in Fig. 5-2(c).

A *cubic* function is one where the highest power of the variable is 3, for example $B(x) = 2x^3 - 8$ or $f(x) = 3x^2 - x^3$. The shape of the graph can be any of the ones illustrated in Fig. 5-2(d).

In general, the graph of a polynomial with highest *degree* (exponent of the variable) n will be a continuous curve that changes direction (between going up and down) at most $n - 1$ times.

SOLVED PROBLEMS

Graphing Functions

1. Use the graph of $y=f(x)$ in Fig. 5-3 to evaluate (a) $f(-3)$, (b) $f(0)$, (c) $f(2)$, and (d) $f(4)$.

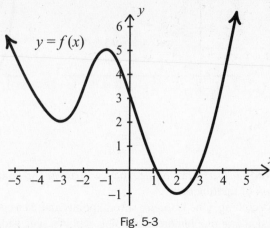

Fig. 5-3

2. If the function graphed in Fig. 5-3 is a polynomial, what is the least possible degree of the function?
3. Graph each of the following functions:

   (a) $f(x)=5$
   (b) $g(x)=2x-3$
   (c) $h(x)=-x+5$
   (d) $k(x)=x^2-2x+1$

   #### Answers

   1. (a) $f(-3)=2$ because the graph goes through the point $(-3, 2)$.
      (b) $f(0)=3$
      (c) $f(2)=-1$
      (d) $f(4)$ is somewhere between 3 and 4. A precise value cannot be read off of a graph unless the coordinates of the point are marked.
   2. The graph of $y=f(x)$ appears to change direction three times: once at $x=-3$, once at $x=-1$, and once at $x=2$. If this function is a polynomial, then it must be of degree at least 4. Thus, this function might be $f(x)=ax^4+bx^3+cx^2+dx+e$ for constants a, b, c, d, and e, but it could not be cubic, quadratic, linear, or constant.
   3. (a) This constant function will graph into the horizontal line consisting of all points with y-coordinate 5, as shown in Fig. 5-4(a).
      (b) The highest power of x in $g(x)=2x-3$ is 1, so the graph will be a straight line. The easiest number to plug into a function is usually $x=0$: the resulting point $(0, -3)$ is called the *y-intercept*. Once a second point is plotted, for example $(2, 1)$, the graph is formed by drawing the straight line through these points, as shown in Fig. 5-4(b).
      (c) The function $h(x)=-x+5$ is also a linear function. To graph it, we plot two points—for example, $(0, 5)$ and $(3, 2)$—and connect them with a straight line, as shown in Fig. 5-5(a).
      (d) This function is quadratic because the highest power of x is 2, thus the graph will be shaped like a parabola. If we plug a few values of x into $k(x)=x^2-2x+1$—for example, $x=0$, 1, 2, and 3—we will get the points $(0, 1)$, $(1, 0)$, $(2, 1)$, and $(3, 4)$. These points make it clear that the parabola turns around at $x=1$. We can thus sketch the graph of the function as in Fig. 5-5(b).

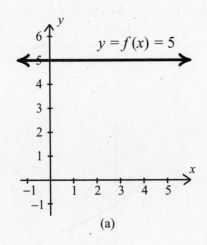

$y = f(x) = 5$

(a)

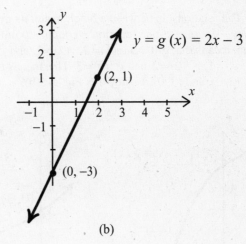

$y = g(x) = 2x - 3$

$(2, 1)$

$(0, -3)$

(b)

Fig. 5-4

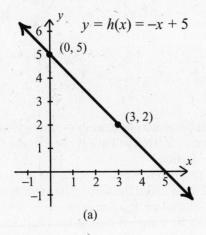

$y = h(x) = -x + 5$

$(0, 5)$

$(3, 2)$

(a)

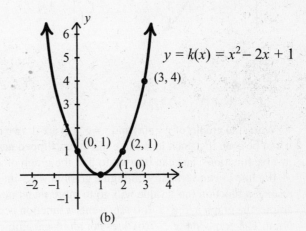

$y = k(x) = x^2 - 2x + 1$

$(3, 4)$

$(0, 1)$ $(2, 1)$

$(1, 0)$

(b)

Fig. 5-5

Inverses to Functions

Every function takes each number in its domain to exactly one number in its range. A *one-to-one* function sends only one number of its domain to each number in its range. For example, the function $f(x) = x^2$ is not one-to-one because the numbers $x = 3$ and $x = -3$ in the domain both go to the same number in the range: 9.

Every one-to-one function f has an *inverse* called f^{-1}, a function that undoes the operation of the function. For example, if $f(1) = 3$, then f takes 1 to 3, so f^{-1} will take 3 back to 1, thus $f^{-1}(3) = 1$. If $f(7) = -11$, then $f^{-1}(-11) = 7$. In general, $f(x) = y$ if and only if $f^{-1}(y) = x$.

This general rule $(f(x) = y \Leftrightarrow f^{-1}(y) = x)$ can be used to find the formula for an inverse function. Set the original function equal to y, then solve for x to find the value of $f^{-1}(y)$.

For example, suppose $f(x) = 5 - 2x$. If $f(x) = y$, then $5 - 2x = y$, so $-2x = y - 5$ and $x = \dfrac{y - 5}{-2}$. Because $f^{-1}(y) = x$, this means that $f^{-1}(y) = \dfrac{y - 5}{-2}$. If you prefer the variable x, plug it in: $f^{-1}(x) = \dfrac{x - 5}{-2}$. We can check that this is an inverse by plugging in points. For example, $f(2) = 5 - 2(2) = 1$ and $f^{-1}(1) = \dfrac{1 - 5}{-2} = \dfrac{-4}{-2} = 2$.

The relationship between a function and its inverse can be seen in their graphs. For example, suppose the graph of a one-to-one function g contains points $(0, 5)$, $(1, 4)$, $(2, 2)$, and $(3, 0)$, as shown in Fig. 5-6(a). This means that $g(0)=5$, $g(1)=4$, $g(2)=2$, and $g(3)=0$. The inverse function g^{-1} must thus take $g^{-1}(5)=0$, $g^{-1}(4)=1$, $g^{-1}(2)=2$, and $g^{-1}(0)=3$. This means that the graph $y=g'(x)$ must go through the points $(5, 0)$, $(4, 1)$, $(2, 2)$, and $(3, 0)$, as shown in Fig. 5-6(b).

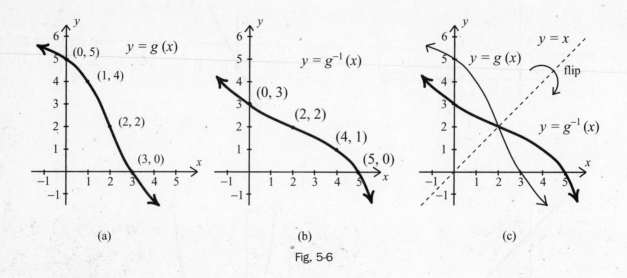

(a)   (b)   (c)

Fig. 5-6

When the graphs of $y=g(x)$ and $y=g^{-1}(x)$ are drawn on the same Cartesian plane, as shown in Fig. 5-6(c), it can be seen that each looks like the other, flipped across the line $y=x$. This will always be the case with inverse functions, and can be used to find the graph of an inverse function.

If a line drawn vertically through a graph always intersects the graph in at most one place, the graph represents a function (no x value will go to more than one y value). This is called the *vertical line test*. For example, the graph in Fig. 5-7(a) represents a function because every dotted vertical line intersects in only one point. The graph in Fig. 5-7(b) does not represent a function because the indicated vertical line intersects the graph in three places.

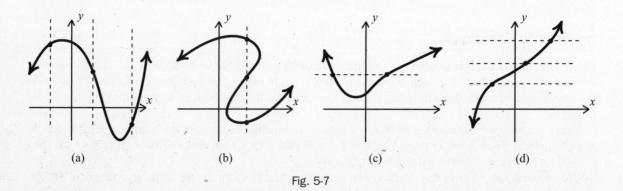

(a)   (b)   (c)   (d)

Fig. 5-7

Similarly, if a line drawn horizontally through the graph of a function only meets in at most one place, then the function is one-to-one (no two x values go to the same y value). This is the *horizontal line test*. For example, the graph in Fig. 5-7(c) is not one-to-one because the indicated horizontal line crosses in two different places. The graph in Fig. 5-7(d) is one-to-one because each horizontal line meets the graph at only one point.

SOLVED PROBLEMS

Inverses to Functions

1. Find the inverse to (a) $f(x) = 4x + 5$, (b) $g(x) = 3 - \frac{1}{2}x$, and (c) $h(x) = \dfrac{x+2}{x-3}$.

2. For each of the graphs in Fig. 5-8, state whether or not it represents a function. If it is a function, state if the function is one-to-one. If the function is one-to-one, sketch the graph of its inverse.

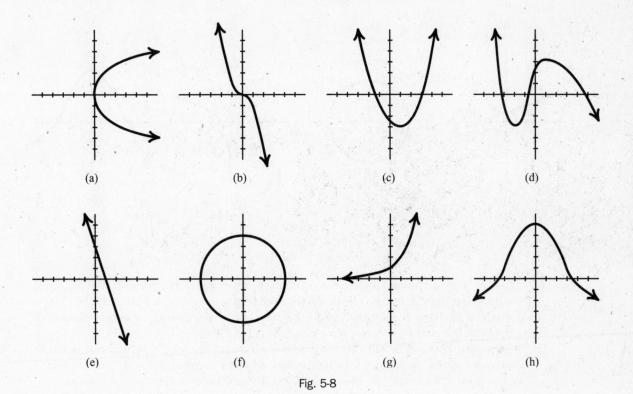

Fig. 5-8

Answers

1. (a) If $f(x) = 4x + 5 = y$, then $4x = y - 5$ and $x = \dfrac{y-5}{4}$. Thus, $f^{-1}(y) = x = \dfrac{y-5}{4}$ and $f^{-1}(x) = \dfrac{x-5}{4}$.

    (b) If $g(x) = 3 - \dfrac{1}{2}x = y$, then $-\dfrac{1}{2}x = y - 3$ and $x = -2y + 6$; thus, $g^{-1}(y) = -2y + 6$ and $g^{-1}(x) = -2x + 6$.

    (c) If $h(x) = \dfrac{x+2}{x-3} = y$, then $x + 2 = (x-3)y = xy - 3y$, so $x - xy = -3y - 2$, $x(1-y) = -3y - 2$, and

    $x = \dfrac{-3y-2}{1-y}$. Thus, $h^{-1}(y) = \dfrac{-3y-2}{1-y}$ and $h^{-1}(x) = \dfrac{-3x-2}{1-x}$.

2. (a) This graph does not represent a function because it fails the vertical line test, as illustrated in Fig. 5-9(a).

    (b) This is the graph of a one-to-one function because it passes both the vertical and horizontal line tests. Its inverse is obtained by reflecting the graph across the line $y = x$, as shown in Fig. 5-9(b). As a guide, it helps to take a few points from the original graph, for example $(-2, 5)$ and $(2, -5)$, to identify points $(5, -2)$ and $(-5, 2)$ on the graph of the inverse.

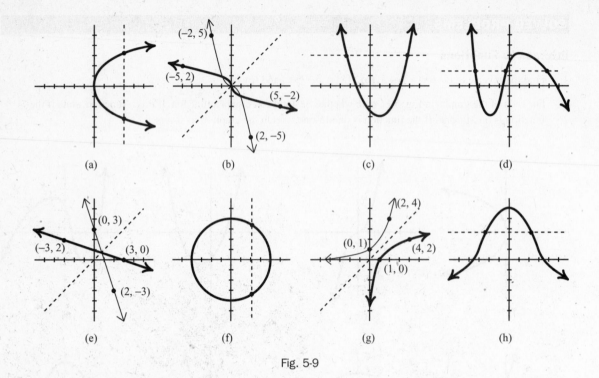

Fig. 5-9

(c) This is the graph of a function because every vertical line crosses it in exactly one place. This function is not one-to-one, however, because a horizontal line can meet it in two places, as illustrated in Fig. 5-9(c).

(d) Just as with (c), this is the graph of a function that is not one-to-one. It passes the vertical line test, but a horizontal line can meet it in three places, as shown in Fig. 5-9(d).

(e) A horizontal or vertical line will meet this straight line in exactly one place; thus, this is the graph of a one-to-one function. We can pick a few points on the graph, for example, (0, 3) and (2, −3), to help find the graph of the inverse, as shown in Fig. 5-9(e). The graph of the inverse will look like the original flipped across the line $y=x$ and go through the points (−3, 2) and (3, 0).

(f) This graph fails the vertical line test, as illustrated in Fig. 5-9(f), and thus does not represent a function.

(g) This graph passes both the horizontal and vertical line tests, and thus represents a one-to-one function. Because the graph goes through the points (0, 1) and (2, 4), the graph of its inverse will go through (1, 0) and (4, 2) and look as though it were flipped across the line $y=x$, as shown in Fig. 5-9(g).

(h) This graph represents a function because it passes the vertical line test. This function is not one-to-one, however, because a horizontal line can cross it in more than one place, as illustrated in Fig. 5-9(h).

Exponential Functions

An *exponent* represents the number of times the *base* is multiplied by itself. For example, 2^5 has a base of 2, has an exponent of 5, and represents the number $2^5 = 2 \cdot 2 \cdot 2 \cdot 2 \cdot 2 = 32$.

A pattern appears when the consecutive powers of a base are listed. In Fig. 5-10, for example, the first six powers of 3 are listed. Each step to the right adds one to the exponent and multiplies the value by 3.

$$3^1 \quad 3^2 \quad 3^3 \quad 3^4 \quad 3^5 \quad 3^6 \quad \cdots$$
$$3 \quad\;\; 9 \quad\;\; 27 \quad\; 81 \quad\;\; 243 \quad 729$$

Fig. 5-10

$$\cdots \quad \begin{array}{cccccccccccc} 3^{-4} & 3^{-3} & 3^{-2} & 3^{-1} & 3^{0} & 3^{1} & 3^{2} & 3^{3} & 3^{4} & 3^{5} & 3^{6} \\ \frac{1}{81} & \frac{1}{27} & \frac{1}{9} & \frac{1}{3} & 1 & 3 & 9 & 27 & 81 & 243 & 729 \end{array} \quad \cdots$$

Fig. 5-11

Similarly, each step to the left subtracts one from the exponent and divides the value by 3. This means that 3^0 is $3^1 = 3$ divided by 3; thus, $3^0 = 1$. Similarly, 3^{-1} is $3^0 = 1$ divided by 3; thus, $3^{-1} = \frac{1}{3}$. Continuing this process, we obtain the list of all integer powers of 3 shown in Fig. 5-11.

In general, any number to the zero power is one. A negative exponent explains how many times the base is divided. For example, $5^{-2} = \frac{1}{5^2} = \frac{1}{25}$.

A function that uses its input variable as the exponent of a base is called an *exponential function*. For example, the exponential function $f(x) = 2^x$ takes 3 to $f(3) = 2^3 = 8$, 4 to $f(4) = 2^4 = 16$, and -3 to $f(-3) = 2^{-3} = \frac{1}{2^3} = \frac{1}{8}$. The graph of $y = f(x) = 2^x$ is illustrated in Fig. 5-12.

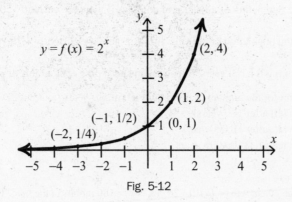

Fig. 5-12

The growth of a population of creatures is one of the best examples of *exponential growth*. Suppose, for example, that there is a forest with 1,000 foxes in it. Suppose further that there is a yearly growth rate of 10%. This means that for every 100 foxes, there will be ten more fox cubs that survive than old foxes that die. At the end of the first year, there will be $1,000 + 0.1 \times 1,000 = 1,100$ foxes. At the end of the second year, there will be $1,100 + 0.1 \times 1,100 = 1,210$. The extra ten fox cubs here are the babies of the previous year's new cubs. Each year there will be more foxes, and thus more babies. After t years, the population of foxes in this forest will be $P(t) = 1,000 \cdot (1 + 0.1)^t$.

In general, a population of A creatures that increases by $k\%$ each year for t years will end up numbering

$$P(t) = A \cdot \left(1 + \frac{k}{100}\right)^t.$$

Instead of knowing the yearly rate of increase, sometimes you will know the time it takes to double (or triple, etc.) in population. If there are A creatures and every n years the population is multiplied by M, then after t years there will be $P(t) = A \cdot M^{\frac{t}{n}}$. For example, if a colony of insects triples in size every two years and begins with 100 insects, then after t years there will be $I(t) = 100 \cdot 3^{\frac{t}{2}}$.

Radioactive materials decrease in quantity over time as they break down, spreading radiation. This is called *radioactive decay*. The *half-life* of a nuclear material is the length of time it takes for half of the material to break down. If you have A units of a radioactive material that has a half-life of n years, then after t years you will have $R(t) = A \cdot \left(\dfrac{1}{2}\right)^{\frac{t}{n}}$ This formula is just the same as the one for

population growth except that every n years we multiply by $\dfrac{1}{2}$ instead of by M.

Money left in the bank will also increase exponentially, as is explained in Chapter 8.

SOLVED PROBLEMS

Exponential Functions

1. Suppose that after t years, the population of rabbits in a field will be $P(t) = 20 \cdot (1.25)^t$.

    (a) How many rabbits will there be in ten years?
    (b) How many rabbits are there today ($t = 0$)?
    (c) What is the rabbit population's annual rate of growth?
    (d) How many rabbits will there be in 50 years? What might be wrong with this?

2. Write a function for the population of a species of dolphins if there are currently 500 dolphins and they increase by 2% each year.

3. Write the function for the population of deer in a forest if there are currently 1,200 and it takes eight years for them to double.

4. The half-life of plutonium is 24,100 years. Suppose someone sprays 500 grams of plutonium all over your house.

    (a) What is the function for the amount of plutonium left after t years?
    (b) How much will be left after 100 years?
    (c) How much will be left after 10,000 years?
    (d) How much will be left after 100,000 years?

Answers

1. (a) After ten years, there will be $P(10) = 20 \cdot (1.25)^{10} \approx 186$ rabbits in the field. To figure this on a calculator, you will have to press 1.25, then the exponent button (either ^ or y^x), then 10 before multiplying by the 20.

    (b) Today there are $P(0) = 20 \cdot (1.25)^0 = 20$ rabbits. The coefficient of the exponential part will always be the *initial population* like this.

    (c) Comparing this formula to the one with the rate of growth—$P(t) = 20 \cdot (1.25)^t = A \cdot \left(1 + \dfrac{k}{100}\right)^t$—we

    see that $1 + \dfrac{k}{100} = 1.25$, so the rabbits increase by $k\% = 25\%$ each year.

    (d) In 50 years, this formula predicts there will be $P(50) = 20 \cdot (1.25)^{50} \approx 1.4$ million rabbits. This is ridiculous. The formula assumes that the rabbits will continue to grow by 25% every year and makes no consideration for them running out of food or space, or for being attacked by predators.

2. The initial population number is $A = 500$. The growth rate is $k\% = 2\%$. Thus, the population function will be

$$P(t) = A \cdot \left(1 + \frac{k}{100}\right)^t = 500 \cdot \left(1 + \frac{2}{100}\right)^t = 500 \cdot (1.02)^t.$$

3. Here we are given the doubling time and not the yearly growth rate. Our initial population is $A = 1,200$, and we know that every $n = 8$ years the population doubles (is multiplied by $M = 2$). The population function is

    thus $P(t) = A \cdot M^{\frac{t}{n}} = P(t) = 1,200 \cdot 2^{\frac{t}{8}}$.

4. (a) Every 24,100 years, our quantity of 500 grams of plutonium will be multiplied by $\dfrac{1}{2}$. Thus, the amount

    after t years will be $P(t) = A \cdot \left(\dfrac{1}{2}\right)^{\frac{t}{n}} = 500 \cdot \left(\dfrac{1}{2}\right)^{\frac{t}{24,100}}$ grams.

(b) After 100 years, there will be $P(100) = 500 \cdot \left(\dfrac{1}{2}\right)^{\frac{100}{24,100}} \approx 498.56$ grams of plutonium still all over your house.

(c) After 10,000 years, there will be $P(10,000) = 500 \cdot \left(\dfrac{1}{2}\right)^{\frac{10,000}{24,100}} \approx 375$ grams of plutonium left.

(d) After 100,000 years, there will be $P(100,000) = 500 \cdot \left(\dfrac{1}{2}\right)^{\frac{100,000}{24,100}} \approx 28$ grams of plutonium left.

Logarithms

The graph of the function $y=f(x)=2^x$ illustrated in Fig. 5-12 is one-to-one because it passes the horizontal line test. It follows that $f(x)=2^x$ has an inverse, written $f^{-1}(x)=\log_2(x)$ and called the *logarithm base 2*. The graph of $y=\log_2(x)$ is illustrated in Fig. 5-13.

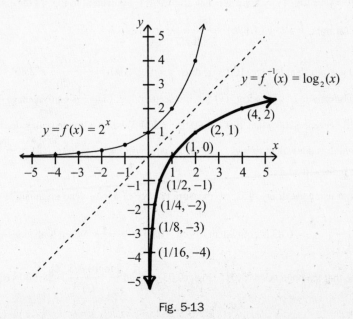

Fig. 5-13

In general, the inverse of $f(x)=b^x$ is the logarithm base b, written $f^{-1}(x)=\log_b(x)$. For example, the inverse of $g(x)=10^x$ is $g^{-1}(x)=\log_{10}(x)$. A base of 10 is assumed whenever a logarithm is written without a base; thus, $\log(x)=\log_{10}(x)$.

Because $f(x)=b^x$ and $f^{-1}(x)=\log_b(x)$ are inverses, the general rule of inverses applies: $f(x)=y \Leftrightarrow f^{-1}(y)=x$. This means that $b^x=y$ is equivalent to $\log_b(y)=x$. This logarithm can be evaluated using the LOG button on a calculator using the change of base formula: $\log_b(y) = \dfrac{\log(y)}{\log(b)}$.

For example, suppose we want to solve $3^x=20$. Using the inverse rule of logarithms ($b^x=y \Leftrightarrow \log_b(y)=x$), with $b=3$, $x=x$, and $y=20$, we see that $3^x=20$ is equivalent to $\log_3(20)=x$. With the change of base formula, $x = \log_3(20) = \dfrac{\log(20)}{\log(3)} \approx 2.7268$. It makes sense that x is between 2 and 3 because $3^2=9<20$ and $3^3=27>20$. Thus, $3^{2.7268} \approx 20$.

SOLVED PROBLEMS

Logarithms

1. Solve for x:

    (a) $2^x = 4{,}096$
    (b) $5^x = 100$
    (c) $\log_4(x) = 3$

2. If \$100 is put in a bank account with a 5% interest rate, the money will become $P(t) = 100 \cdot (1 + 0.05)^t$ after t years. How long will it take for the money to double?

3. Carbon 14 (abbreviated C_{14}) is a rare form of carbon that is radioactive and has a half-life of 5,715 years. All living creatures have a uniform proportion of C_{14} which decays exponentially after they die. Suppose an animal bone is found that has only 3% of the usual amount of C_{14} left. How old is the bone? (Hint: solve for t in $\left(\dfrac{1}{2}\right)^{\frac{t}{5{,}715}} = 0.03$.)

Answers

1. (a) Using $b^x = y \Leftrightarrow \log_b(y) = x$, we see that $2^x = 4{,}096$ is equivalent to $\log_2(4{,}096) = x$. With the change of base formula, $x = \log_2(4{,}096) = \dfrac{\log(4{,}096)}{\log(2)} = 12$.

    (b) $5^x = 100$ is equivalent to $\log_5(100) = x$; thus, $x = \log_5(100) = \dfrac{\log(100)}{\log(5)} \approx 2.86$.

    (c) Using the formula $b^x = y \Leftrightarrow \log_b(y) = x$, we see that $\log_4(x) = 3$ is equivalent to $4^3 = x$; thus, $x = 64$.

2. The \$100 will become \$200 when $100 \cdot (1 + 0.05)^t = 200$. When we divide both sides by 100, we get $(1.05)^t = 2$, which is equivalent to $\log_{1.05}(2) = t$. Thus, it will take $t = \log_{1.05}(2) = \dfrac{\log(2)}{\log(1.05)} \approx 14.2$ years for the money to double.

3. This process of using the decay of C_{14} to measure the age of a long-dead organism is called *carbon dating*. In this case, $\left(\dfrac{1}{2}\right)^{\frac{t}{5{,}715}} = 0.03$ is equivalent to $\log_{\frac{1}{2}}(0.03) = \dfrac{t}{5{,}715}$. When both sides are multiplied by 5,715, we see that the bone is $t = 5{,}715 \cdot \log_{\frac{1}{2}}(0.03) = 5{,}715 \dfrac{\log(0.03)}{\log\left(\dfrac{1}{2}\right)} \approx 28{,}900$ years old.

Logarithmic Scales

Some properties vary tremendously in quantity, for example the strengths of earthquakes, the loudness of sounds, and the lengths of time periods. To compare these, people often use *logarithmic scales*, which represent the quantities by exponents of a common base.

For example, the numbers on the *Richter scale* (used to measure earthquakes) represent exponents of the base 10. A 3 on the Richter scale represents $10^3 = 1{,}000$, the measure of the slightest earth tremor a person could feel. A 5 on the Richter scale represents $10^5 = 100{,}000$, an earthquake 100 times more powerful, which would be generally capable of knocking objects off shelves. An earthquake measuring 9 would be four units more powerful than a 5, thus $10^4 = 10{,}000$ times more powerful, and would be capable of leveling an entire region. In general, every increase of 1 on the Richter scale represents an earthquake that is ten times more powerful.

The *decibel scale* (used to measure sounds) uses a base of $10^{\frac{1}{10}}$ This means that a normal conversation at 60 decibels represents the number $\left(10^{\frac{1}{10}}\right)^{60} = 10^{\frac{60}{10}} = 10^6 = 1,000,000$. The noise of a jet plane taking off is around 120 decibels, representing the number $\left(10^{\frac{1}{10}}\right)^{120} = 10^{\frac{120}{10}} = 10^{12} = 1,000,000,000,000.$ a million times louder. In general, each increase of 10 on the decibel scale represents a sound that is ten times louder.

SOLVED PROBLEMS

Logarithmic Scales

1. How much more powerful is an 8 on the Richter scale than a 7?
2. How much more powerful is a 7 on the Richter scale than a 4?
3. How much more powerful is an 8.7 on the Richter scale than a 7.9?
4. How much louder is an 80-decibel sound than a 60-decibel sound?
5. How much louder is a 105-decibel sound than a 72-decibel sound?

Answers

1. An 8 on the Richter scale is one unit more than a 7, and thus is ten times more powerful.
2. A 7 on the Richter scale is three units more than a 4, and thus is $10^3 = 1,000$ times more powerful.
3. An 8.7 on the Richter scale is 0.8 units higher than a 7.9, and thus is $10^{0.8} \approx 6.3$ times more powerful.

4. An 80-decibel sound is 20 units higher than a 60-decibel sound, and thus is $\left(10^{\frac{1}{10}}\right)^{20} = 10^{\frac{20}{10}} = 10^2 = 100$ times louder.

5. A 105-decibel sound is 33 units higher than a 72-decibel sound, and thus is $\left(10^{\frac{1}{10}}\right)^{33} = 10^{\frac{33}{10}} \approx 1,995$ times louder.

SUPPLEMENTAL PROBLEMS

1. Suppose $f(x) = x^2 + 3x + 2$, $u(t) = 4\sqrt{t}$, and $M(x) = \dfrac{x+5}{x^2-4}$. Evaluate the following:

   (a) $f(3)$
   (b) $u(100)$
   (c) $M(3)$
   (d) $f(-1)$
   (e) $f(t)$
   (f) $u(x)$
   (g) $f(x+h)$
   (h) $M(f(x))$

2. Use the graph of $y=f(x)$ illustrated in Fig. 5-14 to evaluate the following: (a) $f(-1)$, (b) $f(0)$, (c) $f(1)$, and (d) $f(2)$.

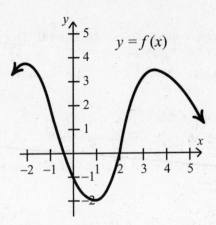

Fig. 5-14

3.   Graph the following:

(a) $y=f(x)=-2$
(b) $y=g(x)=1$
(c) $y = k(x) = \dfrac{1}{2}x + 1$
(d) $y=h(x)=6-2x$
(e) $y=m(x)=x^2-4$
(f) $y=n(x)=x^2-3x+2$

4.   Find the inverse of the following:

(a) $f(x)=8x+10$
(b) $g(x)=4-5x$
(c) $h(x)=x^3+1$
(d) $j(x) = \dfrac{x+1}{2x+5}$

5.   For each of the graphs in Fig. 5-15, state whether it represents a function and, if so, whether that function is one-to-one. Sketch the graph of the inverse to each one-to-one function.

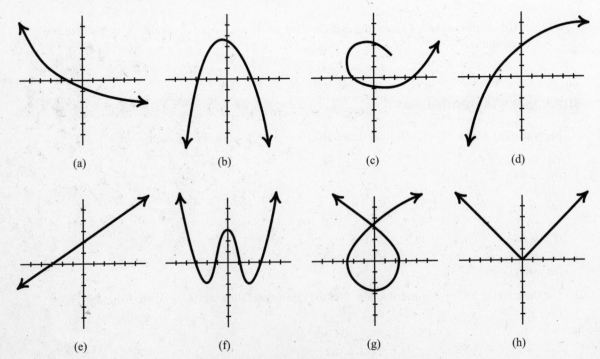

Fig. 5-15

6. If $f(t) = 10 \cdot 3^{\frac{t}{5}}$ then what are (a) $f(15)$, (b) $f(0)$, (c) $f(8)$, and (d) $f(40)$?

7. Every week, an addict of the deadly Substance D requires 20% more of the drug. Bob currently uses 5 grams of Substance D each day.

   (a) Write a function to represent the amount of Substance D Bob will need each day after t weeks.
   (b) What will Bob's daily Substance D habit be in 25 weeks?

8. The man who invented chess asked to be paid in rice. He wanted 1 grain of rice on the first square of the chessboard. Each additional square should have twice as many grains as the last one (two on the second square, four on the third square, etc.).

   (a) Write a function for the amount of rice on the nth square of the board.
   (b) How many grains of rice would be on the last (64th) square?

9. Solve for x:

   (a) $3^x = 70$
   (b) $10^x = 160$
   (c) $2^x = \sqrt{2}$
   (d) $\log_5(x) = 2$

10. There are currently eight mice in a barn. Suppose they triple in number every four months.

    (a) Write a function for the number of mice in the barn after t months.
    (b) How many mice will there be in six months?
    (c) When will there be 100 mice?

11. Suppose 2 grams of a highly radioactive material with a half-life of 100 days is spilled.

    (a) Write a function for the amount of material that will be left after t days.
    (b) How much of the material will be left after a year?
    (c) When will there be only 0.01 grams left?

12. Suppose $10,000 is put in an investment which earns 12% each year.

    (a) Write a function for the value of the investment after t years.
    (b) How long will it take for the money to double?
    (c) How long will it take for the money to triple?

13. Suppose three fossils have (a) 23%, (b) 8%, and (c) 0.01% of the usual amount of carbon 14 (half-life 5,715 years). How old is each fossil?

14. Two earthquakes rock a tiny island. The first measures 8.1 on the Richter scale, and the second measures 7.2. How many times more powerful was the first earthquake?

15. A town bylaw prohibits noises above 70 decibels at night. The sound of a factory at night hits 75 decibels. How many times louder than the acceptable limit is this?

    Answers

    1. (a) 20
       (b) 40
       (c) $\dfrac{8}{5}$
       (d) 0
       (e) $f(t) = t^2 + 3t + 2$
       (f) $u(x) = 4\sqrt{x}$
       (g) $f(x+h) = (x+h)^2 + 3(x+h) + 2$
       (h) $M(f(x)) = \dfrac{f(x)+5}{(f(x))^2 - 4} = \dfrac{x^2 + 3x + 7}{(x^2 + 3x + 2)^2 - 4}$

2. (a) $f(-1)=2$
    (b) $f(0)=-1$
    (c) $f(1)=-2$
    (d) $f(2)=0$

3. Shown in Fig. 5-16.

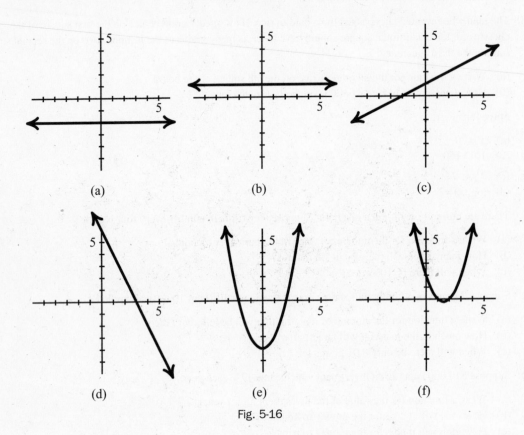

(a)   (b)   (c)

(d)   (e)   (f)

Fig. 5-16

4. (a) $f^{-1}(x) = \dfrac{x-10}{8}$

    (b) $g^{-1}(x) = \dfrac{x-4}{-5}$

    (c) $h^{-1}(x) = \sqrt[3]{x-1}$

    (d) $j^{-1}(x) = \dfrac{5x-1}{1-2x}$

5. (a) A one-to-one function, with the inverse graphed in Fig. 5-17(a)
    (b) A function that is not one-to-one
    (c) Not a function
    (d) A one-to-one function, with the inverse graphed in Fig. 5-17(d)
    (e) A one-to-one function, with the inverse graphed in Fig. 5-17(e)
    (f) A function that is not one-to-one
    (g) Not a function
    (h) A function that is not one-to-one

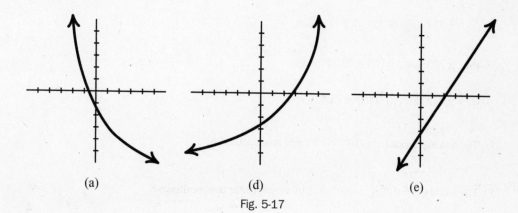

(a)   (d)   (e)

Fig. 5-17

6. (a) 270, (b) 10, (c) $10 \cdot 3^{\frac{8}{5}} \approx 58$, and (d) 65,610

7. (a) $D(t) = 5 \cdot (1.2)^t$
   (b) $D(25) = 5 \cdot (1.2)^{25} \approx 477$ grams

8. (a) $R(n) = 2^{n-1}$
   (b) $R(64) = 2^{63} \approx 9.22 \times 10^{18} = 9,220,000,000,000,000,000$. (Note: when the ruler calculated how much rice this was, the inventor of chess was put to death.)

9. (a) $x = \log_3(70) = \dfrac{\log(70)}{\log(3)} \approx 3.867$
   (b) $x = \log_{10}(160) \approx 2.2$
   (c) $x = \log_2\left(\sqrt{2}\right) = \dfrac{\log\left(\sqrt{2}\right)}{\log(2)} = 0.5$
   (d) $x = 5^2 = 25$

10. (a) $P(t) = 8 \cdot (3)^{\frac{t}{4}}$
    (b) $P(6) = 8 \cdot 3^{\frac{6}{4}} \approx 42$ mice
    (c) $t = 4 \cdot \log_3\left(\dfrac{100}{8}\right) \approx 9.2$ months

11. (a) $P(t) = 2 \cdot \left(\dfrac{1}{2}\right)^{\frac{t}{100}}$
    (b) $2 \cdot \left(\dfrac{1}{2}\right)^{\frac{365}{100}} \approx .16$ grams left
    (c) $t = 100 \cdot \log_{\frac{1}{2}}\left(\dfrac{0.01}{2}\right) \approx 764$ days

12. (a) $P(t) = 10,000 \cdot (1.12)^t$
    (b) $t = \log_{1.12}(2) \approx 6.11$ years
    (c) $t = \log_{1.12}(3) \approx 9.7$ years

13. (a) $5,715 \cdot \log_{\frac{1}{2}}(0.23) \approx 12,117$ years

    (b) $5,715 \cdot \log_{\frac{1}{2}}(0.08) \approx 20,825$ years

    (c) $5,715 \cdot \log_{\frac{1}{2}}(0.0001) \approx 75,939$ years

14. The first earthquake was $10^{0.9} \approx 7.9$ times more powerful.

15. The factory is $\left(10^{\frac{1}{10}}\right)^5 = 10^{\frac{5}{10}} \approx 3.16$ times louder than is allowed.

CHAPTER 6

Geometry

The word *geometry* means "earth measurement" because it started with surveying (measuring land). Every year, for example, the Nile River would flood and erase all boundary markings on the world's most fertile ground. It was essential for the harmony of the ancient Egyptian civilization that ownership and responsibility over various plots of land were reestablished as soon as the waters receded. To this effect, surveyors would stretch ropes, mark off distances, and draw lines in the mud to designate each person's lot. Over time, the study of lines, areas, angles, and volumes proved useful in many other areas.

Lengths

The most basic geometric measurement is *length*, the distance along a line. For example, the height of a person, the distance between cities, and the width of a river are all measured by length.

The *perimeter* of an object is the distance around it. For a *polygon*, a figure enclosed by straight sides, the perimeter is found by adding up the lengths of the sides. For example, the perimeter of a rectangle with a base of length b and a height of h is $2b + 2h$, as illustrated in Fig. 6-1.

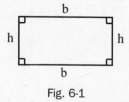

Fig. 6-1

The distance across a circle is called the *diameter*, as illustrated in Fig. 6-2. Half of this, the distance from the center to the edge, is called the *radius*. The perimeter of a circle is called the *circumference*. A piece of string wrapped around a circular object will always be a little more than three diameters in length. This curious ratio of circumference C to diameter D is approximately 3.14159265, usually represented by the Greek letter π (pi). Thus, $\dfrac{C}{D} = \pi$, so $C = \pi \cdot D$ or $C = 2\pi \cdot r$ because $D = 2 \cdot r$, where r is the radius. For most calculations, $\pi \approx 3.14$ is sufficient.

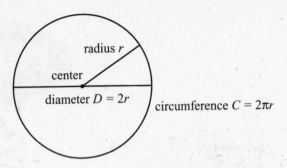

Fig. 6-2

SOLVED PROBLEMS

Lengths

1. Find the perimeter of each figure in Fig. 6-3.

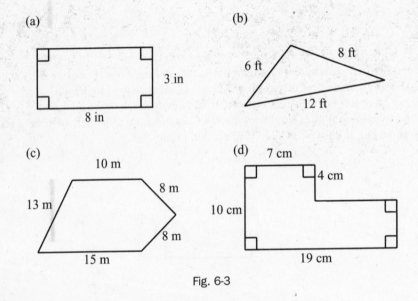

(a)

3 in

8 in

(b)

6 ft

8 ft

12 ft

(c)

10 m

8 m

13 m

8 m

15 m

(d)

7 cm

4 cm

10 cm

19 cm

Fig. 6-3

2. For each circle in Fig. 6-4, find the diameter, radius, and circumference.

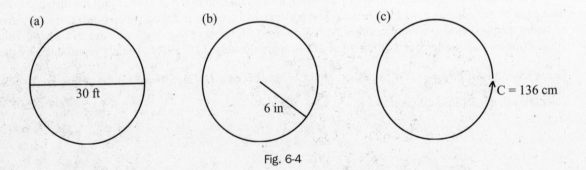

(a)

30 ft

(b)

6 in

(c)

C = 136 cm

Fig. 6-4

3. Suppose an athletic field has the dimensions given in Fig. 6-5. How many laps are required to run one kilometer? (Hint: one kilometer is 1,000 meters.)

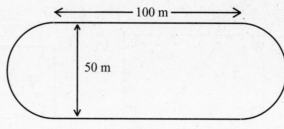

Fig. 6-5

Answers

1. (a) perimeter $= 8+3+8+3 = 22$ in
    (b) perimeter $= 6+8+12 = 26$ ft
    (c) perimeter $= 13+10+8+8+15 = 54$ m
    (d) Here the perimeter $= 10+7+4+12+6+19 = 58$ cm, the same as if it were a 10 cm by 19 cm rectangle, as illustrated in Fig. 6-6.

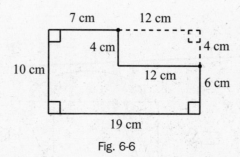

Fig. 6-6

2. (a) $D = 30$ ft, $r = 15$ ft, and $C = 30\pi \approx 30(3.14) = 94.2$ ft
    (b) $r = 6$ in, $D = 12$ in, and $C = 12\pi \approx 12(3.14) = 37.68$ in
    (c) $C = 136$ cm $= \pi D$, so $D = \dfrac{136}{\pi} \approx \dfrac{136}{3.14} \approx 43.312$ cm and $r \approx 21.656$ cm

3. The two ends form a circle of radius 25 m, which thus has a circumference of $2\pi \cdot (25) \approx 50 \cdot (3.14) = 157$ m. Together with the two straight parts, a full lap consists of 357 meters. A kilometer thus consists of $\dfrac{1000}{357} \approx 2.8$ laps.

Areas

The next most basic geometric measurement is *area*. Even in ancient times, people knew that the amount of grain expected of a harvest could be calculated using the area of the field. Many objects are measured by area, for example rugs, windows, walls, and driveways.

The area of a rectangle is either base times height ($A = b \times h$) or length times width ($A = l \times w$), depending on what its two dimensions are called, as shown in Fig. 6-7(a).

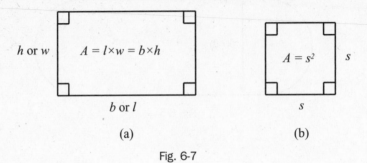

(a)   (b)

Fig. 6-7

A square is a rectangle with both dimensions the same; thus, a square with each side of length s will have area $A = s \times s = s^2$, as shown in Fig. 6-7(b). It is because of this that multiplying a number by itself is called *squaring* the number.

A *parallelogram* is a four-sided figure where each pair of opposite sides is parallel. Lines are *parallel* if they would never intersect, even if extended straight forever. If a parallelogram has a base length b and height h, then its area will be $A = b \times h$. This is because its area can be cut and rearranged into a $b \times h$ rectangle, as illustrated in Fig. 6-8.

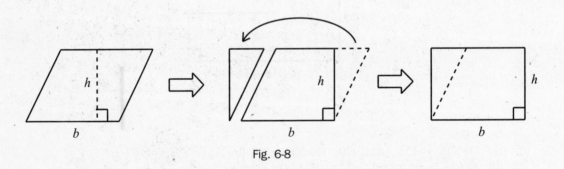

Fig. 6-8

The area of a triangle with base b and height h is $A = \frac{1}{2} b \cdot h$. This is because the area is half that of a parallelogram with base b and height h, as illustrated in Fig. 6-9.

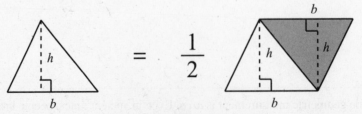

Fig. 6-9

A *trapezoid* is a four-sided figure with exactly one pair of parallel sides. A trapezoid with parallel sides of length a and b a distance of h apart has half the area of an $(a+b)$ by h parallelogram, as illustrated in Fig. 6-10. This means that the area is $\frac{1}{2}(a+b)\cdot h$.

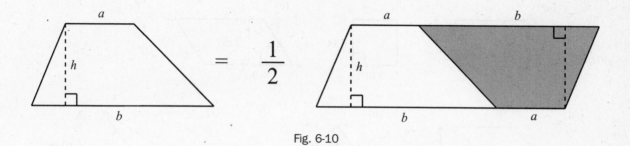

Fig. 6-10

If the area of a circle is broken into a great number of *sectors* (wedges from the center), it can be re-arranged into an approximate parallelogram, as illustrated in Fig. 6-11. When the sectors are very small, the top and bottom will be very close to flat. The height of the parallelogram will be r, the radius of the circle. The circumference of the circle will form the top and the bottom of the rectangle, so the width of the parallelogram will be $\frac{1}{2}C = \pi \cdot r$. This leads us to conjecture that the area of a circle is $(\pi \cdot r) \cdot r = \pi \cdot r^2$.

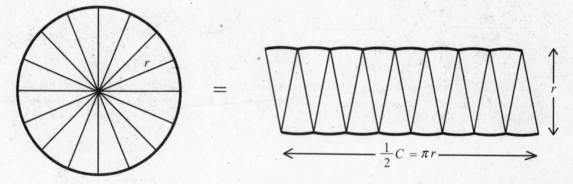

Fig. 6-11

SOLVED PROBLEMS

Areas

1. Find the area of each shape in Fig. 6-12.

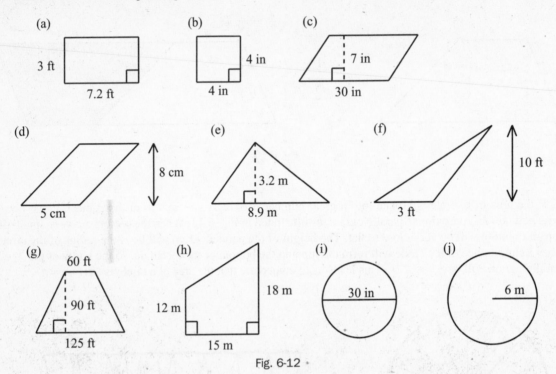

Fig. 6-12

2. Suppose a carpet-cleaning service charges 10 cents per square foot. How much will it cost to clean the rug illustrated in Fig. 6-13(a)? How much will it cost to clean the wall-to-wall carpeting in the room whose floor plan is illustrated in Fig. 6-13(b)?

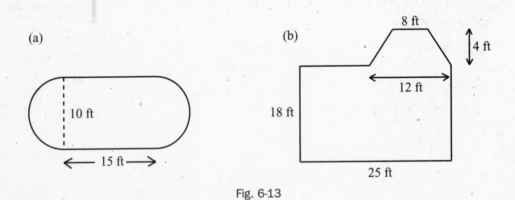

Fig. 6-13

Answers

1. (a) This rectangle has area $= l \times w = 3 \cdot (7.2) = 21.6 \, ft^2$.
    (b) This square has area $= s^2 = 4^2 = 16 \, in^2$.
    (c) This parallelogram has area $= b \times h = 7 \times 30 = 210 \, in^2$.
    (d) This parallelogram has area $= b \times h = 5 \times 8 = 40 \, cm^2$.

    (e) This triangle has area $= \frac{1}{2} b \cdot h = \frac{1}{2}(8.9)(3.2) = 14.24 \, m^2$.

(f)　This triangle has area $= \frac{1}{2}b \cdot h = \frac{1}{2}(3)(10) = 15\,ft^2$.

(g)　This trapezoid has area $= \frac{1}{2}(a+b) \cdot h = \frac{1}{2}(125+60) \cdot 90 = 8,325\,ft^2$.

(h)　This is a trapezoid with area $= \frac{1}{2}(a+b) \cdot h = \frac{1}{2}(12+18) \cdot 15 = 225\,m^2$.

(i)　The radius of this circle is $r = 15$ in, so the area $= \pi r^2 = \pi \cdot (15)^2 \approx (3.14) \cdot 225 = 706.5\,in^2$.

(j)　This circle has area $= \pi r^2 = \pi \cdot (6)^2 \approx (3.14) \cdot 36 = 113.04\,m^2$.

2.　The area of the throw rug can be broken into a 15×10 foot rectangle (area $= 15 \times 10 = 150\,ft^2$) and a circle with a ten-foot diameter (radius $= 5$ ft). The total area is approximately $150 + (3.14) \cdot 5^2 \approx 228.5\,ft^2$, so it should cost $228.5 \times (0.10) = \$22.85$ to clean.

The area of the room can be broken into a rectangle and a trapezoid. The rectangle has area $18 \times 25 = 450\,ft^2$. The trapezoid has area $\frac{1}{2}(8+12) \cdot 4 = 40\,ft^2$. Thus, the cost to clean the floor should be $(450 + 40) \times (0.10) = \$49$.

Volumes

Volume measures the amount of three-dimensional space inside an object. For example, the storage space in a trailer, the amount of liquid in a bottle, and the amount of air in a balloon are all measured by volume.

It is easy to calculate the volumes of *prisms*, objects where the top, bottom, and every horizontal cross-section have the same shape and area. The volume of a prism is obtained by multiplying the height by the area of the base.

For example, a box is a prism because the top and bottom are identical rectangles. Suppose a box has length l, width w, and height h, as illustrated in Fig. 6-14(a). The area of the base is $l \times w$, so the volume is $l \times w \times h$. Sometimes a box is called a *rectangular prism* because its base is a rectangle.

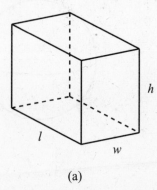　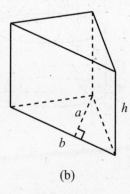　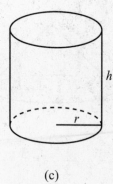

(a)　　　　　(b)　　　　　(c)

Fig. 6-14

A prism with a triangle for a base is called a *triangular prism*. Suppose a triangular prism has height h and the base triangle has base b and altitude a (*altitude* is another name for height), as illustrated in Fig. 6-14(b). The area of the base is $\frac{1}{2}b \cdot a$, so the volume of the triangular prism is $\frac{1}{2}b \cdot a \cdot h$.

A prism with a circle for a base is a *cylinder*. Suppose a cylinder has height h and a base circle with radius r, as shown in Fig. 6-14(c). The area of the base is πr^2, so the volume of the cylinder is $\pi r^2 h$.

When a three-dimensional object tapers to a point from a flat base, its volume is one-third that of a prism with the same height and base. For example, the pyramid in Fig. 6-15(a) has a height of h, an s by s square base, and a volume of $\frac{1}{3}s^2h$. Similarly, the cone with height h and base radius r in Fig. 6-15(b) has volume $\frac{1}{3}\pi r^2 h$.

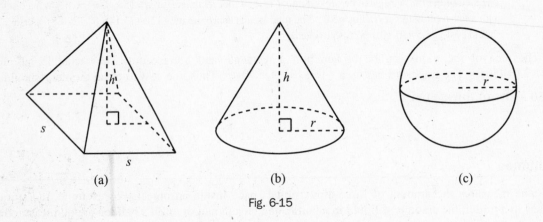

(a)   (b)   (c)

Fig. 6-15

The volume of a sphere with radius r, as shown in Fig. 6-15(c), has volume $\frac{4}{3}\pi r^3$. This leads to a curious result (illustrated in Fig. 6-16): the volume of a sphere with radius r is the same as the combined volumes of a cone and a cylinder, both with radius r and height r. This was first discovered by Archimedes around 250 B.C.

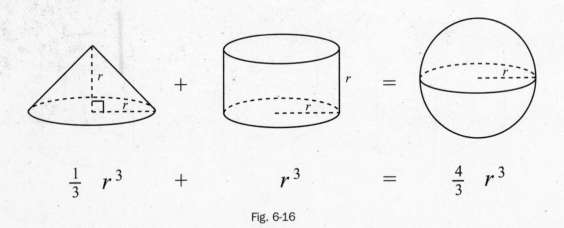

$$\frac{1}{3}\ r^3\qquad +\qquad r^3\qquad =\qquad \frac{4}{3}\ r^3$$

Fig. 6-16

SOLVED PROBLEMS

Volumes

1.   Find the volume of each object illustrated in Fig. 6-17.

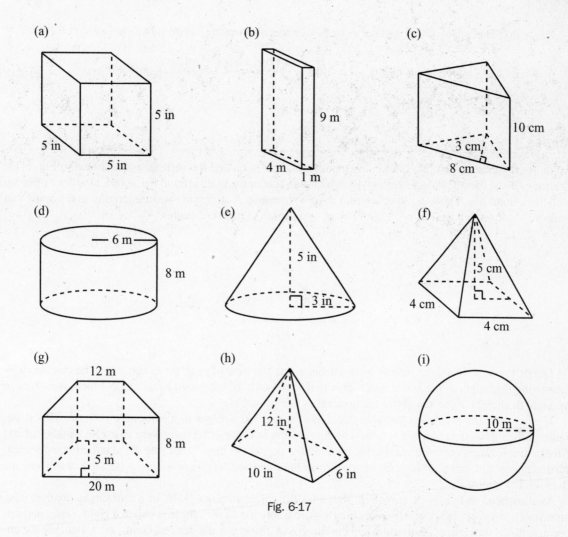

(a)

5 in
5 in
5 in

(b)

9 m
4 m
1 m

(c)

10 cm
3 cm
8 cm

(d)

6 m
8 m

(e)

5 in
3 in

(f)

5 cm
4 cm
4 cm

(g)

12 m
8 m
5 m
20 m

(h)

12 in
10 in
6 in

(i)

10 m

Fig. 6-17

Answers

1. This cube has volume $5 \times 5 \times 5 = 5^3 = 125\,\text{ft}^3$. Incidentally, this is why raising a number to the third power is called *cubing*.

    (b) This box has volume $1 \times 4 \times 9 = 36\,\text{m}^3$.

    (c) The area of the base of this triangular prism is $\frac{1}{2}(3)(8) = 12\,\text{cm}^2$;   thus, the volume is $12 \times 10 = 120\,\text{cm}^3$.

    (d) The volume of this cylinder is $\pi(6)^2 \cdot 8 = 288\pi \approx 288 \cdot (3.14) = 904.32\,\text{m}^3$.

    (e) The volume of this cone is $\frac{1}{3}\pi(3)^2 \cdot 5 = 15\pi \approx 15 \cdot (3.14) + 47.1\,\text{in}^3$.

    (f) The volume of this square-bottomed pyramid is $\frac{1}{3}(4)^2 \cdot 5 = \dfrac{80}{3} = 26\frac{2}{3}\,\text{cm}^3$.

    (g) The base of this prism is a trapezoid with area $\frac{1}{2}(12 + 20) \cdot 5 = 80\,m^2$, so the volume is $80 \times 8 = 640\,\text{m}^3$.

(h)   The base of this pyramid is a rectangle with area $6 \times 10 = 60 \text{ in}^2$, so the volume is $\frac{1}{3}(60) \cdot 12 = 240 \text{ in}^3$.

i.   The volume of this sphere is $\frac{4}{3}\pi \cdot (10)^3 = \dfrac{4000\pi}{3} \approx \dfrac{4000 \cdot (3,14)}{3} = 4{,}186\frac{2}{3}\, m^3$.

Angles

An *angle* is formed when two lines come together at a point (called the *vertex*), as illustrated in Fig. 6-18. The measure of the angle is the amount by which one line must rotate around the vertex in order to line up with the second line. This is a rather unusual thing to measure. A stick can measure lengths, and a bottle can measure volumes, but there is no natural object with which to measure angles.

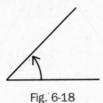

Fig. 6-18

The first people to measure angles were astronomers. The time of year, for example, can be calculated by measuring the height of the sun at noon. This height can only be measured as an angle—there is no way to measure the height of the sun above the horizon in feet or inches.

The astronomers of ancient Babylon decided to count 360 *degrees* in a full-circle rotation. This is because of their base-sixty number system. The Babylonian astronomical charts were so useful that the ancient Greeks and Romans copied them. In this manner, the degree (written °) became the standard unit for measuring angles. Similarly, we have 60 minutes in an hour and 60 seconds in a minute because of the way the Babylonians counted.

As illustrated in Fig. 6-19, there are 360° in a full-circle rotation, 180° in a half-circle rotation (also known as a *straight angle*), and 90° in a quarter-circle rotation. A 90° angle is called a *right angle*, perhaps because this is the right angle to use when building most objects (walls, fences, boxes, etc.). Because the angles of squares are right, 90° angles are often indicated by a small square in the corner. Lines that meet at right angles are called *perpendicular*.

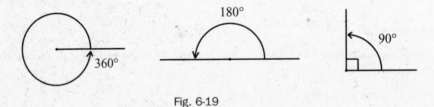

Fig. 6-19

Angles which are less than 90° are called *acute angles*, for example those in Fig. 6-20(a) and 6-20(b), and angles which are between 90° and 180° are called *obtuse angles*, for example the angles in Fig. 6-20(c) and 6-20(d).

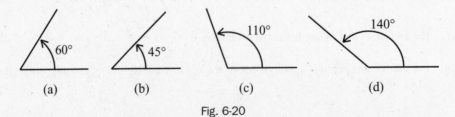

Fig. 6-20

Two straight lines of the same length are called *congruent*. A triangle with two congruent sides is called *isosceles*. The angles opposite the congruent sides will have the same measure (called *congruent angles*), as shown in Fig. 6-21(a). The converse of this is also true: if two angles of a triangle are congruent, then the sides opposite will have the same length. This is illustrated in Fig. 6-21(b).

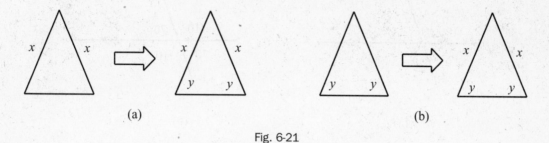

(a)   (b)

Fig. 6-21

When two straight lines cross, four angles with measures a, b, c, and d are formed, as illustrated in Fig. 6-22(a). The angles with measure a and b together form a straight angle; thus, $a+b=180°$. Such a pair of angles is called *supplementary*. With algebra, we see that $a=180°-b$. Similarly, $b+c=180°$, so $c=180°-b$. It follows that $a=c$. Angles which are on the opposite sides of an intersection like this are called *vertical angles* and will always have the same measure. Angles with the same measure are called *congruent*. One example is illustrated in Fig. 6-22(b). Any two angles formed by an intersection are either vertical and congruent (the 30° pair or the 150° pair) or else *adjacent* (next to each other) and supplementary.

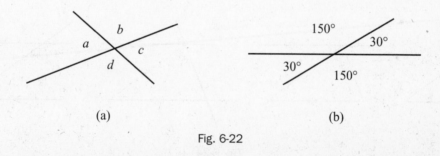

(a)   (b)

Fig. 6-22

When a line *transverses* (runs across) a pair of parallel lines, many of the angles formed will be congruent. For example, the two horizontal lines in Fig. 6-23(a) are parallel (indicated by the arrows) and crossed by a diagonal transversal. The angles marked x and y are called *alternate interior angles* because they are between the two parallel lines (in the interior) and on alternate sides of the transversal. Assuming that everything is flat (as we will for the remainder of this section), these angles will be congruent: $x=y$. If $x=70°$, we get the picture in Fig. 6-23(b) when the measures of all the vertical and supplementary angles are calculated. This pattern will always appear when parallel lines are crossed by a transversal.

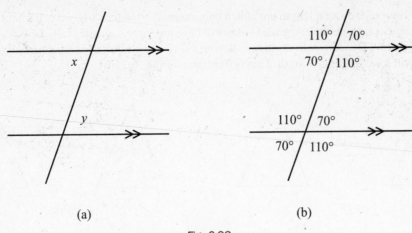

(a)   (b)

Fig. 6-23

If the measures of the angles of a triangle are a, b, and c, as illustrated in Fig. 6-24(a), then $a+b+c=180°$. To prove this, draw a line parallel to the base through the top vertex, as shown in Fig. 6-24(b). The left side of the triangle can be extended to transverse these parallel lines; thus, the two angles marked a in Fig. 6-24(c) are congruent alternate interior angles. Similarly, the two angles marked c in Fig. 6-24(d) are congruent because they are interior angles on alternate sides of the right side of the triangle. Because the angles of measure a, b, and c form a straight angle across the top, they sum as follows: $a+b+c=180°$.

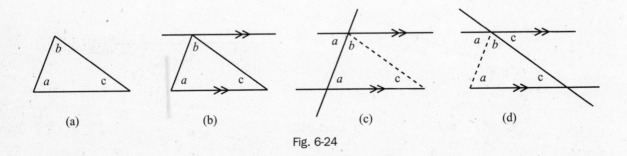

(a)   (b)   (c)   (d)

Fig. 6-24

A triangle with all three sides congruent is called *equilateral*. Because it is isosceles in every way possible, all of its angles must be congruent. In order for these three congruent angles to sum to 180°, each must be 60°.

SOLVED PROBLEMS

Angles

1. Find the measure of each marked angle and side in Fig. 6-25.

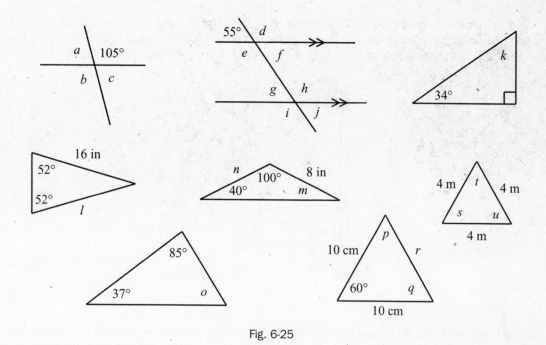

Fig. 6-25

Answers

1.   Given in Fig. 6-26.

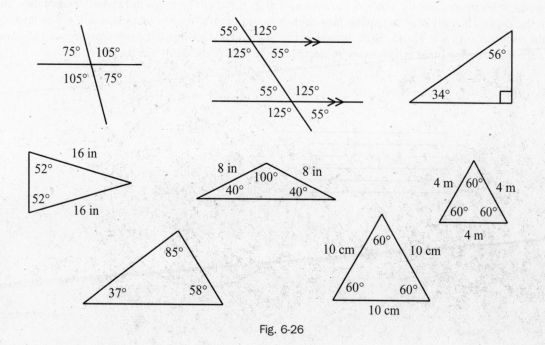

Fig. 6-26

How Eratosthenes Measured the Earth around 250 B.C.

With the geometry detailed in the previous sections, we can understand how Eratosthenes, the head of the Library of Alexandria in Egypt, was able to measure the earth in 250 B.C.

Around that time, Eratosthenes found a curious letter from Syene, a town in southern Egypt. The letter said that at noon on the summer solstice (June 21), the sun was directly overhead. The writer knew this because sunlight could be seen reflecting up from the water in a deep well (illustrated in Fig. 6-27(a)).

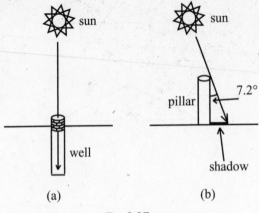

Fig. 6-27

Eratosthenes was intrigued by this account. On the next summer solstice, he watched the shadow of a pillar as noon approached. If the sun went exactly overhead, he reasoned, the shadow would disappear entirely. This did not happen. The shadow grew short before noon, but then began to lengthen. Eratosthenes marked the shortest length of the shadow and used it to calculate that the sun was about one-50th of a full circle (about 7.2°) away from directly overhead at noon, as illustrated in Fig. 6-27(b).

Eratosthenes figured that the sun was so far away that its rays must be essentially parallel when they fall upon the earth. The only way to explain how the sun could be overhead in one place and not in another would be if the earth was round, as illustrated in Fig. 6-28. (The fact that the earth casts a round shadow upon the moon during lunar eclipses was another reason to believe the earth was round.)

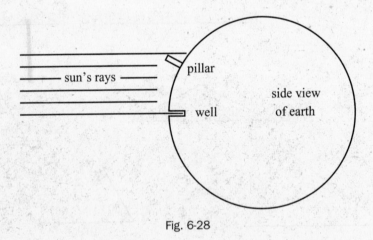

Fig. 6-28

Furthermore, he reasoned that if the well and the pillar were both perfectly vertical to the ground, then they must both point at the center of the earth. Even better, the angle at which these lines would meet at the center of the earth must be an alternate interior angle to the one Eratosthenes measured, and thus must be 7.2°, as illustrated in Fig. 6-29.

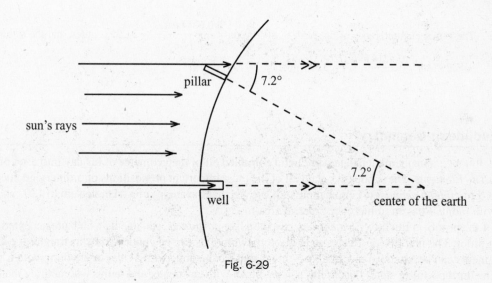

Fig. 6-29

Because 7.2° is one-50th of a full circle, this meant that the distance from the pillar in the city of Alexandria to the well in the town of Syene was one-50th of the complete distance around the earth. Fortunately, the distances between some cities in Egypt had already been measured: the distance between Alexandria and Syene was 5,000 *stades* (a unit of measurement equal to the standard distance around a sports stadium). It followed that the circumference of the earth was 50×5,000=250,000 stades. If these measurements use the standard distance of 157 meters around an Egyptian stadium (measured by archaeologists unearthing ancient arenas), then Eratosthenes estimated the circumference of the earth at about 250,000×157=39,250,000 m=39,250 km. This is remarkably close to the earth's actual circumference of about 40,100 km.

SOLVED PROBLEMS

Eratosthenes

1. Some scholars insist that Eratosthenes would have used the Greek stade, the distance around an Olympic stadium, which was 185 meters. If so, what would have been his estimate for the circumference of the earth?
2. What would the radius of the earth be using (a) the Egyptian stade, (b) the Olympic stade, and (c) the actual circumference of the earth?
3. What are the actual radius and circumference of the earth in miles? Use 0.6214 miles per kilometer.

Answers

1. $250,000 \; stades \times \dfrac{185 \, m}{1 \, stade} = 46,250,000 \; m = 46,250 \, km$

2. Because $C = 2\pi r, \; r = \dfrac{C}{2\pi} \approx \dfrac{C}{2 \cdot (3.14)}$. Thus,

    (a) $r \approx \dfrac{39,250}{2 \cdot (3.14)} = 6,250 \, km$

    (b) $r \approx \dfrac{46,250}{2 \cdot (3.14)} \approx 7,365 \, km$

    (c) $r \approx \dfrac{40,100}{2 \cdot (3.14)} \approx 6,385 \, km$

3. The actual circumference of the earth is $40,100\,km \times \dfrac{0.6214\,miles}{1\,km} \approx 24,918$ miles, and the actual radius

is $6,385\,km \times \dfrac{0.6214\,miles}{1\,km} \approx 3,968$ miles.

Non-Euclidean Geometry

Around 300 B.C., a mathematician named Euclid compiled all of the geometry of his day into a set of books entitled *The Elements*. These books contain all of the most important propositions of ordinary geometry, often called *Euclidean geometry*. For example, it is proved that the angles of a triangle sum to 180° and that a circle with radius r has circumference $2\pi r$ and area πr^2.

All of the proofs in Euclid's *Elements* are based on five *postulates*, statements which are accepted as true without proof. The first two postulates state that any two points can be connected by a straight line and that line segments can be extended indefinitely—things that can be done with a ruler or anything with a straight edge. The third postulate states that circles can be drawn with any center and radius—something that can be done with a compass or a piece of string. The fourth postulate states that all right angles have the same measure—something that a square or a protractor can measure. Thus, the first four of Euclid's postulates lay out the basic tools of geometry: the compass, the straight edge, and the square. Most people have readily accepted that these postulates are true.

Euclid's fifth postulate is equivalent to the following: if there is a point not on a line, then there is exactly one line through the point which is parallel to the line. This is illustrated in Fig. 6-30.

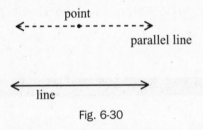

Fig. 6-30

Most mathematicians felt Euclid made this a postulate only because he did not know how to prove it. They attempted to improve *The Elements* by finding a proof. This was classically known as the *fifth postulate problem*. However, no one was able to prove it without introducing new postulates which could not be proven.

Eventually people realized that there were different kinds of geometry. Only in Euclidean geometry is the fifth postulate true. On one hand, if we change the fifth postulate to say that there are no parallel lines, we end up with *elliptic geometry*. On the other hand, if we suppose that a point not on a line has many different parallel lines through it, we will end up with *hyperbolic geometry*.

Elliptic geometry is geometry as performed on a very large sphere, which is natural because *geometry* means "earth measurement" and the earth is a sphere. If a line is extended straight forever on a sphere, it will eventually loop back around to form a *great circle*, as illustrated in Fig. 6-31(a). Every great circle, like the equator, divides the sphere into two equal-sized hemispheres. Any other straight line would form another great circle, and thus could not be entirely above or below the first line (this would not divide the earth into equal hemispheres). Thus any two lines must intersect in two places, as shown in Fig. 6-31(b). It is for this reason that there are no parallel lines in elliptic geometry.

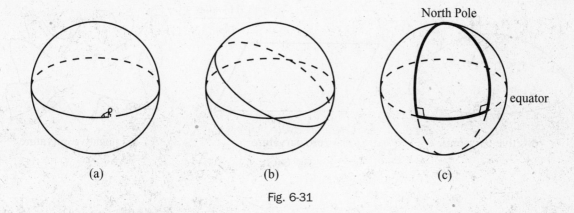

Fig. 6-31

Without parallel lines, we cannot prove that the angles of a triangle sum to 180°. Instead, the angles of a triangle on a sphere will always sum to more than 180°. For example, a triangle formed by the equator and two lines of longitude through the North Pole will have two right angles, as illustrated in Fig. 6-31(c). Furthermore, the sum of the angles will depend on the size of the triangle: big triangles (for example, the one in Fig. 6-31(c)) will have an angle sum much larger than 180°, while small triangles will have an angle sum just barely larger than 180°. This means that there are no similar triangles of different sizes in elliptic geometry.

Circles are also different on the spheres of elliptic geometry than they are on the flat planes of Euclidean geometry. For example, the lines of latitude are all circles centered at the North Pole, as illustrated in Fig. 6-32(a). Notice that even though the radius r_1 from the North Pole to the equator is smaller than the radius r_2 from the North Pole to a latitude in the southern hemisphere, the circumference of the equator is greater than that of the further latitude, as shown in Fig. 6-32(b) and 6-32(c). In general, a circle with radius r on a sphere will have a circumference less than $2\pi r$ and an area less than πr^2.

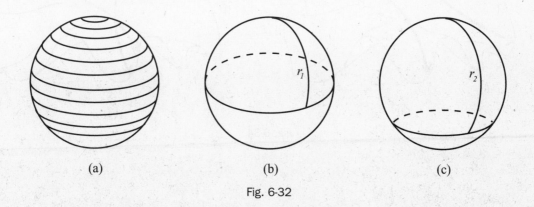

Fig. 6-32

Euclidean geometry is based on flat planes which are said to have *zero curvature*. Elliptic geometry takes place on planes that are *positively curved* into giant spheres. The third kind of geometry, hyperbolic geometry, takes place on planes which are *negatively curved* and frilly like certain kinds of seaweed. These three kinds of curvature are illustrated in Fig. 6-33.

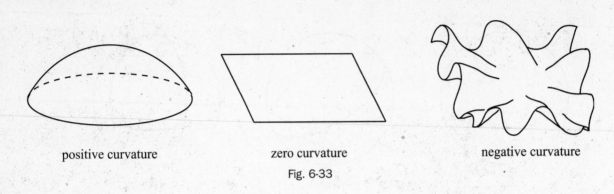

positive curvature   zero curvature   negative curvature

Fig. 6-33

Just as with elliptic space, any tiny piece of hyperbolic space will be almost flat and have properties very close to those of Euclidean space. As pieces get larger, however, they contain more of the warps, bends, and frills that are symptomatic of this space. The word *hyperbolic*, by the way, has roots meaning "excessive" and "exaggerated."

Figure 6-34(a) illustrates a point *P* which is not on the line through points *A* and *B*. Because of all the ripples and folds in this hyperbolic space, there are many lines which can be drawn straight through *P* that will never intersect with the line through *A* and *B*. This shows how hyperbolic space satisfies the alternative to Euclid's fifth postulate which results in many parallel lines.

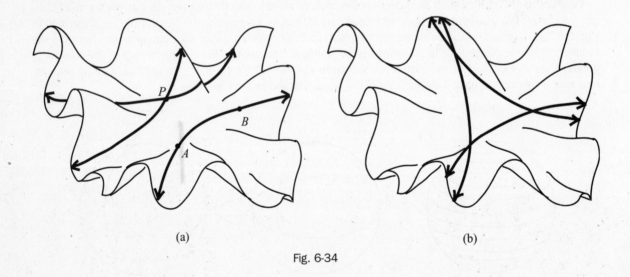

(a)   (b)

Fig. 6-34

Figure 6-34(b) shows how three straight lines can form a triangle in hyperbolic space with three very small angles, for an angle sum of less than 180°. Small triangles will have an angle sum close to 180°, while large triangles will have a much smaller angle sum. Just as with elliptic space, this means that there are no similar triangles of different sizes in hyperbolic space.

Because of all the curves and ripples in hyperbolic space, a circle of radius *r* will have circumference greater than $2\pi r$ and area greater than πr^2.

For comparison purposes, the properties of the three kinds of geometry are detailed in Table 6-1.

TABLE 6-1

	ELLIPTIC GEOMETRY	EUCLIDEAN GEOMETRY	HYPERBOLIC GEOMETRY
Lines	Finite great circles	Infinite straight lines	Infinite straight lines
Planes	Finite positively curved spheres	Infinite flat planes of zero curvature	Infinite negatively curved planes
Angle sum of a triangle	$> 180°$	$=180°$	$< 180°$
Similar triangles of different sizes?	No	Yes	No
Circumference of a circle with radius r	$< 2\pi r$	$=2\pi r$	$> 2\pi r$
Area of a circle with radius r	$< \pi r^2$	$=\pi r^2$	$>\pi r^2$
Number of lines parallel to a given line through a point	0 (no parallel lines)	Exactly one	Infinitely many

Unsolved Problems: Non-Euclidean Geometry

1. If you look up into the nighttime sky, can you imagine two lines that run parallel into space, extending forever but never intersecting? Can you even mentally picture a single line that is infinite in length? Are you inclined to believe that planes in our universe are curved like spheres, infinite and flat like Euclidean planes, or infinite and negatively curved?
2. Different versions of the fifth postulate, believed without proof, lead to very different sorts of geometry. What other sorts of things do people believe without proof? Could these beliefs shape the nature of their lives?

Higher Dimensions

Through most of history, the study of geometry was limited to objects that could be seen and drawn. The ancient Greeks, for instance, insisted that every proof was accompanied by a figure that could be drawn only with a compass and straight edge. Even infinitely long lines are a relatively recent concept; Euclid used finite line segments which were only extended when necessary.

The next big step toward the abstraction of geometry was made by René Descartes in 1637. It was already known that points on a line can be identified with numbers; for example, the point 1.3 on the number line is illustrated in Fig. 6-35(a). The Cartesian plane identified each point on a two-dimensional plane with a pair of real numbers. As an example, the point $(4, -2)$ is illustrated in Fig. 6-35(b).

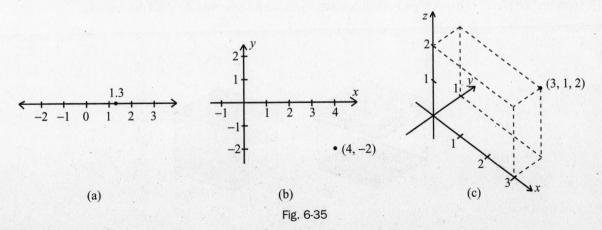

(a)　　　　　　　(b)　　　　　　　(c)

Fig. 6-35

Similarly, every point in three-dimensional space can be described by a triple of real numbers (x, y, z). The z coordinate gives the distance up and away from the x-y plane. This point can be viewed as the furthest tip of a box with one vertex at the origin and sides of length x, y, and z. For example, the point $(3, 1, 2)$ is illustrated in Fig. 6-35(c).

In 1854, Bernhard Riemann argued that points in higher-dimensional spaces can be represented by larger numbers of coordinates. A point in four-dimensional space, for example, can be described using four coordinates: (x, y, z, w). An ordered set of five coordinates, for example $(2, 7, -3, 1, 9)$, describes a point in five-dimensional space. The fact that such spaces can be impossible to illustrate should not stop people from using them.

As an example, the computers in modern cars record information that is often useful after an accident or a breakdown. A car that keeps track of the time, car speed, engine speed, engine temperature, fuel-air mix, and position of the brake is recording six coordinates at a time. As the car drives, this data forms points in six-dimensional space. A mechanic can use the information without having to visualize the space.

For mathematicians, it can be entertaining and educational to try to visualize higher-dimensional spaces. The most useful approach is to look for patterns among the more familiar dimensions.

Dimensions can be viewed as directions, ways in which things can change or travel. If there are no dimensions, there are no ways a point can move or change. Thus, zero-dimensional space consists of a single point, as illustrated in Fig. 6-36(a).

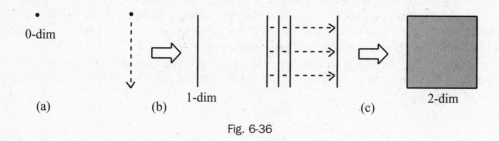

Fig. 6-36

If that single point is given a dimension, it will be able to move back and forth in that single direction. If the point moves for a unit of distance in this direction, then the set of all the points through which it has traveled will form a one-dimensional line segment, as illustrated in Fig. 6-36(b).

With a second dimension, the line segment will be able to move in a direction perpendicular to the first. Moved one unit, the result will be a two-dimensional square, as illustrated in Fig. 6-36(c).

If that square is moved in a third dimension for one unit, the result will be a three-dimensional cube, as illustrated in Fig. 6-37. Notice that as the square moves down, it appears to overlap itself. This is because the page is only two-dimensional. As we are familiar with three dimensions, we can easily imagine that these stack up like pieces of paper (they have different heights—coordinates in the third dimension).

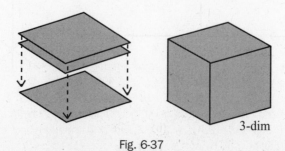

Fig. 6-37

Following this pattern, if our three-dimensional cube is moved one unit in a fourth dimension, the result will be a four-dimensional cube (also called a *hypercube*), as illustrated in Fig. 6-38. Even though it looks as though the cubes overlap, they do not because they have different coordinates in the fourth dimension. Physicists often use time as a fourth dimension. In this case, the boxes do not overlap because they exist at different times.

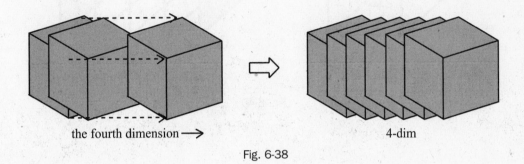

Fig. 6-38

Another way to envision the fourth dimension is to contemplate empty containers in various dimensions. Even though we live in a three-dimensional world, we generally use three dimensions to describe empty objects, for example the space in a room or the volume of a bottle. With solid three-dimensional objects, for example human beings, we only look upon their surfaces and try not to think about all the sloshy stuff inside them. Actually, if time is considered a fourth dimension, then human beings are four-dimensional creatures (our fourth lengths are the lengths of our lives).

To best contemplate a container, it helps to imagine something inside that is trying to get out. Thus, we will design prisons for the worst criminal in each dimension, assuming that each one can move freely in his or her space.

For a point in zero-dimensional space, however evil it may be, there is no need to build a prison. It already has no freedom of movement, as illustrated in Fig. 6-39(a).

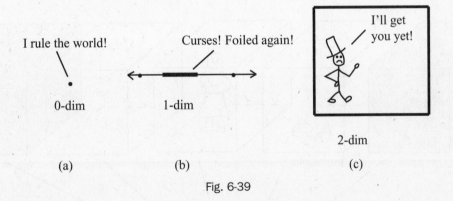

Fig. 6-39

If there were a one-dimensional villain, for example a maniacal line segment, it could be kept away from the rest of the line with two points, as illustrated in Fig. 6-39(b).

A two-dimensional criminal can be enclosed in a square, as illustrated in Fig. 6-39(c).

A wicked three-dimensional creature can be contained in a box, as illustrated in Fig. 6-40(a). Notice that every vertical slice through this box will be a hollow square (a two-dimensional container) except for the two ends which are solid squares. The two ends and a few example slices are illustrated in Fig. 6-40(b).

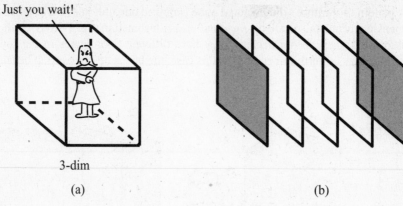

Just you wait!

3-dim

(a)   (b)

Fig. 6-40

Now suppose that there is an evil being capable of traveling freely in four dimensions. If we use time as the fourth dimension, this means that she can not only fly but also travel through time. We will suppose she can only go from one time to the next by going through all the times in between. If she could disappear and reappear in a completely different place or time, then it would be impossible to contain her.

One way to trap this four-dimensional being would be to build a box with a small hole in the top and bottom. First, stop up the bottom and fill the box with sand using the hole in the top. Next, open up the bottom and drain out the box. At this point, the time-traveling thief can be placed into the box, along with some provisions. She will probably imagine that this is merely a three-dimensional box and that she can escape through the fourth dimension. Either she will travel back in time to before the box was built and escape, or else she will travel forward in time until the box crumbles with age. To foil her, we fill the box back up with sand. Before she is smothered completely, the thief will escape back in time to before it was filled. If she goes much further back in time, she will discover that the box had been filled with sand earlier. She is therefore trapped in a four-dimensional container, as illustrated in Fig. 6-41.

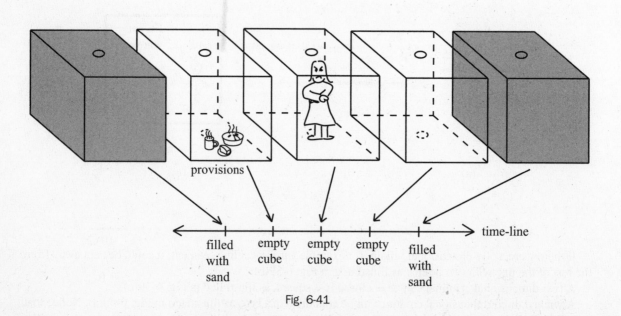

provisions

filled with sand   empty cube   empty cube   empty cube   filled with sand   time-line

Fig. 6-41

SOLVED PROBLEMS

Higher Dimensions

1. Suppose a weather device records the time, temperature, air pressure, wind speed, and humidity. How many dimensions are there to this data?
2. Suppose an engineer builds a robot that has 20 joints. The robot's computer keeps track of the position of each joint, from 0 (all the way closed) to 100 (all the way open). How many dimensions are there to the set of all possible positions for the robot?
3. A one-dimensional container is formed by two zero-dimensional points. A two-dimensional container (a square) is formed by four one-dimensional line segments. A three-dimensional container (a cube) is formed by six two-dimensional squares. What can thus be concluded about four- and five-dimensional containers?
4. How many vertices are there on (a) a zero-dimensional object, (b) a one-dimensional line segment, (c) a two-dimensional square, (d) a three-dimensional cube, (e) a four-dimensional hypercube, and (f) a five-dimensional cube?

Answers

1. This data has five dimensions.
2. Each of the robot's positions can be represented by a point in 20-dimensional space. Some of the positions could be impossible: for example, the limbs cannot overlap. Other positions could involve the robot losing its balance. This means that moving a robot from one position to another can require moving around unusable points in 20-dimensional space.
3. A four-dimensional cube is formed by eight solid three-dimensional cubes. A five-dimensional cube is formed by ten solid four-dimensional hypercubes. Each container is formed by objects from the previous dimension, and their number follows the following pattern: 2, 4, 6, 8, 10 …
4. (a) There is one vertex on a zero-dimensional object.
    (b) There are two endpoint vertices on a one-dimensional line segment.
    (c) There are four corner vertices on a two-dimensional square.
    (d) There are eight corner vertices on a three-dimensional cube.
    (e) So far, a container in each dimension has twice as many vertices as the previous dimension. Thus, we suppose there are 16 vertices on a four-dimensional hypercube.
    (f) There are 32 vertices on a five-dimensional cube.

SUPPLEMENTAL PROBLEMS

1. Find the perimeter of each shape in Fig. 6-42.

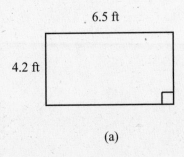

6.5 ft

4.2 ft

(a)

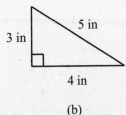

5 in

3 in

4 in

(b)

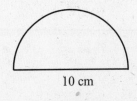

10 cm

(c)

Fig. 6-42

2.   Find the diameter d, radius r, circumference C, and area A of a circle with (a) $d=8\,$m, (b) $r=3\,$cm, (c) $C=37.2\,$ft, and (d) $A=200\,in^2$.

3.   Find the area of each object in Fig. 6-43.

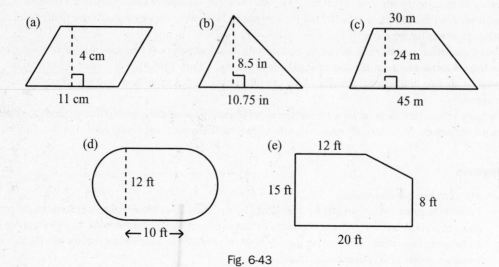

Fig. 6-43

4.   The windows of a house are 55 inches tall and 30 inches wide. A box of window insulation contains a sheet of plastic that is 62 inches by 210 inches. How many windows can it cover?

5.   Suppose a spherical weather balloon has a diameter of 12 feet. How many cubic feet of helium will be necessary to fill the balloon?

6.   A small can is 4.5 inches tall and three inches across (diameter). A larger can is 5.5 inches tall and 3.5 inches across. How many times larger is the volume of the second can?

7.   Two customs officers are about to open a wooden box that is 2.5 feet wide, four feet long, 1.7 feet tall, and perhaps filled with poisonous snakes. What is the volume of the box?

8.   Suppose a stack of grain forms a rough cone shape that is four feet tall and 15 feet across. Roughly how many cubic feet of grain is this?

9.   Find the volume of each object in Fig. 6-44.

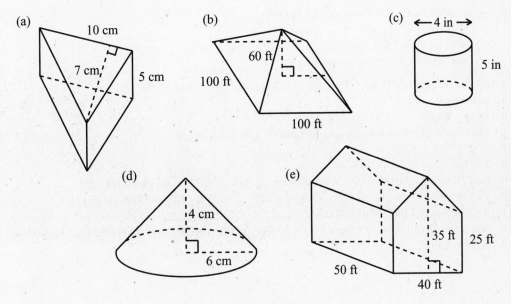

Fig. 6-44

10. Find the measure of each marked side or angle in Fig. 6-45.

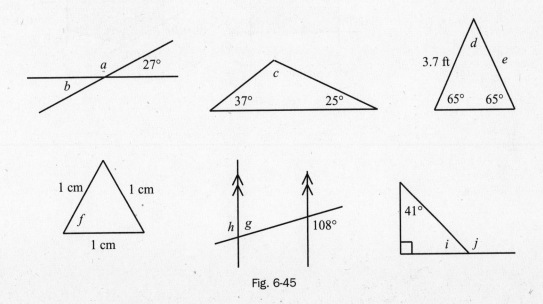

Fig. 6-45

11. If the circumference of the earth is 40,100 km, what is its volume?

12. What sort of non-Euclidean geometry has (a) no parallel lines, (b) no lines of infinite length, (c) circles with radius r and area $> \pi r^2$, (d) a triangle angle sum $> 180°$, (e) negatively curved planes, or (f) no similar triangles of different sizes?

13. Draw a (a) line segment, (b) square, and (c) cube with all sides of length 1, and then find the coordinates for each vertex.

Answers

1. (a) 21.4 ft, (b) 12 in, and (c) $10 + 5\pi$ cm

2. (a) $d=8\,\text{m}$, $r=4\,\text{m}$, $C=8\pi\,\text{m}$, and $A=16\pi\,m^2$
   (b) $d=6\,\text{cm}$, $r=3\,\text{cm}$, $C=6\pi\text{cm}$, and $A=9\pi\,cm^2$
   (c) $d\approx11.8\,\text{ft}$, $r\approx5.9\,\text{ft}$, $C=37.2\,\text{ft}$, and $A\approx109.3\,ft^2$
   (d) $d\approx16\,\text{in}$, $r\approx8\,\text{in}$, $C\approx50\,\text{in}$, and $A=200\,in^2$

3. (a) $44\,cm^2$, (b) $45.6875\,in^2$, (c) $900\,m^2$, (d) $120+36\pi\approx233ft^2$, and (e) $272\,ft^2$
4. Seven windows
5. $288\pi\approx904\,ft^3$
6. The volume of the second can is approximately 1.66 times larger.
7. $17\,ft^3$
8. $235.5\,ft^3$
9. (a) $175\,cm^3$, (b) $200{,}000\,ft^3$, (c) $20\pi\approx62.8\,in^3$, (d) $48\pi\approx150.72\,in^3$, and (e) $60{,}000\,ft^3$
10. $a=153°$, $b=27°$, $c=118°$, $d=50°$, $e=3.7\,\text{ft}$, $f=60°$, $g=72°$, $h=108°$, $i=49°$, and $j=131°$
11. Approximately $1{,}089{,}990{,}000{,}000\,km^3$
12. (a) elliptic, (b) elliptic, (c) hyperbolic, (d) elliptic, (e) hyperbolic, and (f) both elliptic and hyperbolic
13. See Fig. 6-46.

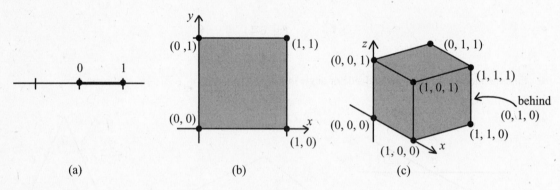

Fig. 6-46

CHAPTER 7

Graph Theory

A diagram made of points connected by lines is called a *graph*, not to be confused with the sorts of graphs formed by plotting x- and y-coordinates on the Cartesian plane. These sorts of graphs are very useful for conveying information and presenting certain problems.

The Bridges of Königsberg

The story of graph theory starts in Königsberg, a city on the Baltic coast, back in the 1700s. This city was built around two islands in the Pregel River, all connected by seven bridges, as illustrated in Fig. 7-1.

The citizens of this Prussian city had an open challenge: find a way to walk through the city in such a way as to cross every bridge exactly once. Many people had a pleasant walk while attempting to solve this puzzle, but no one found a way to do it. Those who thought they had figured it out invariably either had skipped a bridge or else went over one more than once. For example, a person who started at the north shore and walked over bridge 1, then 3, then 4, then 2, then 6, and then 7 would have missed bridge 5 and have no way to get there without crossing over 3, 4, or 7 a second time.

In 1736, the great Swiss mathematician Leonhard Euler (pronounced "oiler") was able to prove that no such walk was possible. To do so, he invented a new sort of mathematics. In general, it is easy to prove that something is possible: just do it. To prove that something cannot be done, however, requires cleverness.

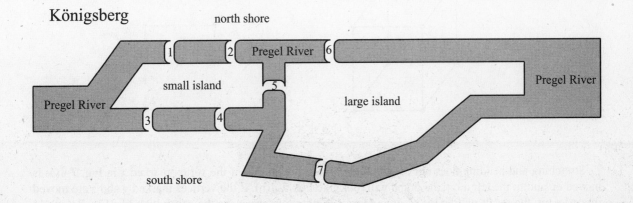

Fig. 7-1

Graphs

Euler's first step was to simplify the map of Königsberg by reducing the four parts of the city to points, connected by lines representing the seven bridges, as shown in Fig. 7-2(a). Any journey over the bridges of the city could be described entirely as a path from point to point over the lines of this figure. In fact, for the purposes of this problem, the map of Königsberg can be reduced to Fig. 7-2(b). The distances between the bridges and the sizes of the islands were irrelevant.

What Euler drew was a *graph*. A graph is a mathematical figure consisting of points called *vertices* which are connected by lines called *edges*. The *star* of a vertex is formed by the first piece of each edge which begins or ends at the vertex. The *valence* or *degree* of a vertex is the number of edge-pieces in its star.

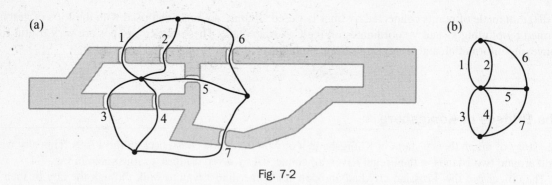

Fig. 7-2

For example, the edges of a cube form a graph with 12 edges and eight vertices, as shown in Fig. 7-3(a). The vertices where the edges meet are represented by large dots to differentiate from where edges appear to cross but do not actually meet. In Fig. 7-3(b), the star of each vertex is emphasized. It can be seen that each star is formed by three edge-pieces; each vertex has degree 3.

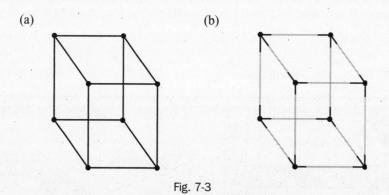

Fig. 7-3

Stretching and bending does not change the graph. For example, if the vertex marked *x* in Fig. 7-4(a) is moved up and to the left a bit, the graph will look like Fig. 7-4(b). If the vertices marked *y* and *z* are moved inward a bit, the result will be Fig. 7-4(c). More straightening can make the graph look like Fig. 7-4(d). In general, it can be very difficult to recognize when two large graphs are the same if they are *presented* (shown) differently. The presentations in Fig. 7-4(c) and 7-4(d) are called *planar* because they are portrayed on a piece of paper (a plane) without edges crossing over one another.

When geometric objects are considered unchanged by stretching and bending, the result is *topology*, a branch of mathematics which studies the innate properties of different kinds of space.

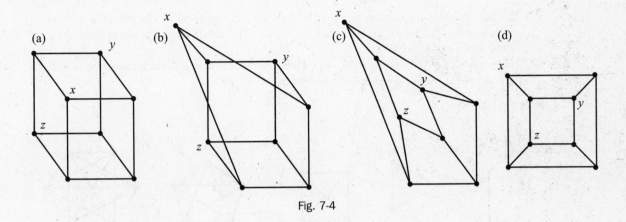

Fig. 7-4

SOLVED PROBLEMS

Graphs

1. For the graphs in Fig. 7-5, give the number of vertices, the number of edges, and the degrees of each vertex.

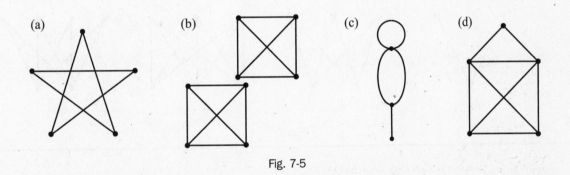

Fig. 7-5

2. The *complete graph* with n vertices, written K_n, has n vertices and one edge between each pair of edges. Draw (a) K_1, (b) K_2, (c) K_3, (d) K_4, (e) K_5, and (f) K_6.

3. The *complete bipartite graph* $K_{m,n}$ has $m+n$ vertices. Each of the m vertices is connected by one edge to each of the n vertices. Draw (a) $K_{2,1}$, (b) $K_{3,2}$, (c) $K_{3,3}$, and (d) $K_{4,2}$.

4. Draw graphs (answers will vary) with the following properties:

   (a) Five edges and three vertices
   (b) Two vertices, each of degree 4
   (c) Three vertices, each of degree 2
   (d) Three vertices, with degrees 1, 2, and 3
   (e) Four vertices, with degrees 1, 2, 3, and 4

5. If the degrees of all the vertices of a graph are added, is it possible for the total to be an odd number?

   ### Answers

   1. (a) This graph has five vertices and five edges. Each vertex has degree 2. A pentagon would be another presentation of this graph.
      (b) This graph has 12 edges and eight vertices, all of degree 3. One big difference between this graph and the one in Fig. 7-3 is that this graph is not *connected*: we cannot travel along edges from any vertex to any other vertex.

(c) This graph has four edges connecting three vertices. The top edge is called a *loop* because it begins and ends at the same vertex. The top vertex has degree 4, the middle vertex has degree 3, and the bottom vertex has degree 1.

(d) This graph has five vertices and eight edges. The top vertex has degree 2. The bottom vertices both have degree 3. The middle vertices both have degree 4.

2. Shown in Fig. 7-6.

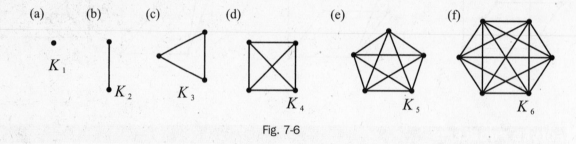

Fig. 7-6

3. Shown in Fig. 7-7.

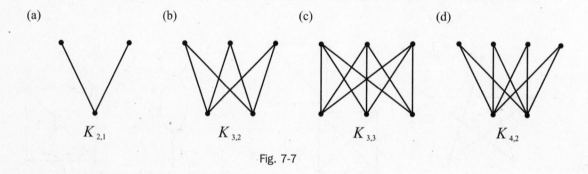

Fig. 7-7

4. Some possible answers are given in Fig. 7-8.

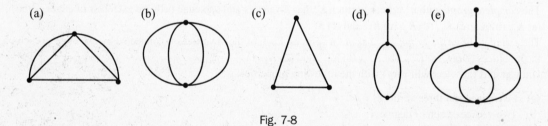

Fig. 7-8

5. Each edge in a graph forms one part of the star for two vertices (or two parts of one star if the edge is a loop). This means that when the degrees of all the vertices in a graph are added up, the total will count all the edges twice. Thus, the sum of the degrees of all the vertices will always be an even number.

Euler Paths and Circuits

A *path* in a graph is a sequence of edges where each one begins where the last one ends. A *circuit* is a path which begins and ends at the same vertex. A path or circuit is *Euler* if it uses every edge in the graph exactly once.

For example, below the graph in Fig. 7-9(a) is a diagram indicating how one could proceed around the edges of the graph. This is a path because each edge begins where the last one ended; you could walk this path. This is Euler because the path goes over every edge exactly once. This not a circuit, however, because the path begins at the lower-left vertex and ends at the lower-right vertex.

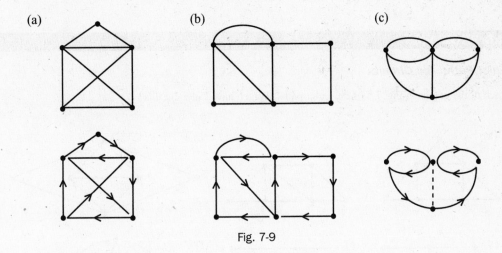

Fig. 7-9

Below the graph in Fig. 7-9(b) is a diagram illustrating an Euler circuit for the graph; every edge is used exactly once, and the path begins and ends at the same vertex.

The diagram below the graph in Fig. 7-9(c) illustrates a path that begins at the left vertex and ends at the right vertex. This is not a circuit. This is also not Euler because the vertical edge in the middle is not used.

The bridges of Königsberg problem can thus be phrased: "Is there an Euler path for the graph in Fig. 7-2(b)?"

When working on this problem, Euler realized that every vertex of an Euler circuit must have an even degree. When a vertex is visited once, for example, there will be a total of two edges connected to it: one with which to go to the vertex and another with which to leave. This is illustrated in Fig. 7-10(a). To visit a vertex a second time will require two more edges, as shown in Fig. 7-10(b). The vertex at which the circuit begins and ends will have two more edges in its star: the first and last edges of the path, as shown in Fig. 7-10(c).

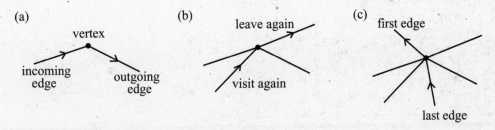

Fig. 7-10

Because "If a graph has an Euler circuit, then every vertex of the graph has even degree" is true, its contrapositive is also true: "If a graph has a vertex of odd degree, then it cannot have an Euler circuit." The converse is also true: "If every vertex of a graph has an even degree, then the graph has an Euler circuit." It is necessary, of course, that the graph be connected. This converse will not be proven here, though it is not difficult to find an Euler circuit for a connected graph with every vertex of even degree.

A connected graph with exactly two odd-degree vertices, for example the graph in Fig. 7-9(a), will not have an Euler circuit but will have an Euler path. The path will begin at one of the odd-degree vertices and end at the other one. A graph with more than two odd-degree vertices, for example the graph in Fig. 7-9(c), will have neither an Euler circuit nor an Euler path.

Because the graph for the bridges of Königsberg has four vertices with odd degrees (three with degree 3 and one with degree 5), it follows that there is neither an Euler circuit nor a path. Euler thus announced to the city of Königsberg that it was absolutely impossible to walk through town and cross over every bridge exactly once.

SOLVED PROBLEMS

Euler Paths and Circuits

1.   Which of the graphs in Fig. 7-11 have Euler paths? Which have Euler circuits?

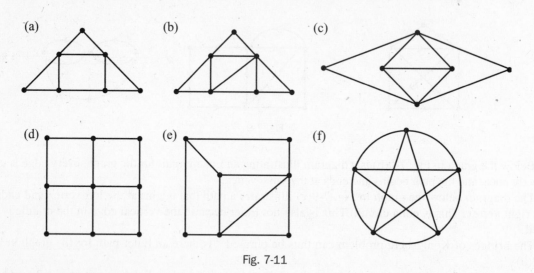

Fig. 7-11

2.   Is a every circuit a path?
3.   Is every path a circuit?
4.   Suppose there is an Euler circuit for the streets of a neighborhood. Of what use might this be for the mail carrier?

Answers

1.   The stars and degrees for all the vertices are given in Fig. 7-12.

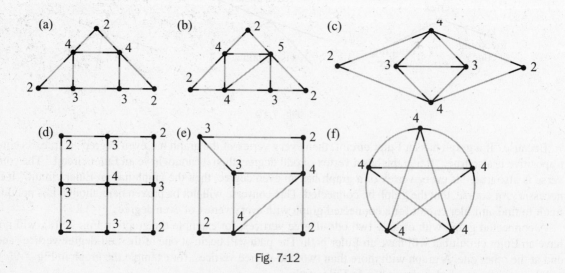

Fig. 7-12

(a) This graph has two odd-degree vertices, and thus has an Euler path but not an Euler circuit. Any Euler path will have to begin at one degree-3 vertex and end at the other.

(b) This graph has two odd-degree vertices, and thus has an Euler path but not an Euler circuit. Each Euler path will begin at either the degree-3 or degree-5 vertex and end at the other.

(c) This graph has two odd-degree vertices, and thus has an Euler path but not an Euler circuit. Each Euler path will have to begin and end at the degree-3 vertices.

(d) This graph has four odd-degree vertices (the four 3's), and thus has neither an Euler circuit nor an Euler path.

(e) This graph has two odd-degree vertices, and thus has an Euler path but not an Euler circuit. Each Euler path will have to begin and end at the two degree-3 vertices.

(f) The vertices of this graph are all of even degree (4); thus, there is an Euler circuit (which is also an Euler path). The circuit can begin and end at any one of the five vertices.

2. A circuit is a path, specifically one that begins and ends at the same vertex. Thus, every circuit is a path.

3. Not every path is a circuit—only those which begin and end at the same vertex.

4. The mail carrier will be able to park the mail truck, deliver the mail for the whole neighborhood, and end up back at the truck without walking down any street more than once.

Hamiltonian Paths and Circuits

In 1856, an Irish mathematician named William Hamilton invented the Icosian game. The goal of this game was to find a circuit that went once though every vertex of a *dodecahedron*, a three-dimensional object formed by 12 identical pentagons, as illustrated in Fig. 7-13(a). The graph formed by the edges of this object is shown in Fig. 7-13(b), and a planar representation is shown in Fig. 7-13(c).

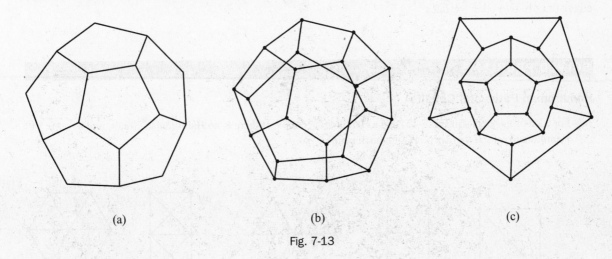

(a)   (b)   (c)

Fig. 7-13

In honor of this mathematician, a path or circuit in a graph which visits every vertex exactly once is called *Hamiltonian*.

For example, the diagram below the graph in Fig. 7-14(a) illustrates a Hamiltonian circuit.

The graph in Fig. 7-14(b) has a Hamiltonian path, as shown in the diagram below it. A Hamiltonian circuit would have to use the two edges highlighted in Fig. 7-14(c) in order to pass through the upper-left vertex. Similarly, all eight of the edges highlighted in Fig. 7-14(d) would need to be used to go through all four corner vertices. This circuit cannot be extended to include the middle vertex without using one of the side vertices twice, and thus this graph has no Hamiltonian circuit.

The graph in Fig. 7-14(e) does not even have a Hamiltonian path. Every edge connects the center vertex to one of the outer vertices. When one of these is chosen, as shown in Fig. 7-14(f), we can continue on to

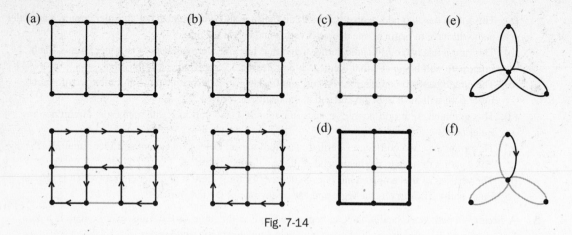

Fig. 7-14

one of the remaining two vertices, but not both. A path that goes through all four vertices must thus go through the center more than once.

It is interesting to note that Fig. 7-14(a) has a Hamiltonian circuit, but not even an Euler path. The graph in Fig. 7-14(e) has an Euler circuit, but no Hamiltonian path. In general, there is no connection between a graph having an Euler path or circuit and having a Hamiltonian path or circuit.

In 1952, a mathematician named Paul Dirac came up with a useful theorem about Hamiltonian circuits. In a graph with $n \geq 3$ vertices, there will always be a Hamiltonian circuit so long as every vertex is connected by edges to at least $\frac{n}{2}$ other vertices (the degree of each vertex must be $\geq \frac{n}{2}$, not counting loops or multiple edges between the same vertices). This theorem can be used to guarantee that a graph has a Hamiltonian circuit, but cannot be used to prove that a graph does not have a Hamiltonian circuit. The graph in Fig. 7-14(a), for example, has 12 vertices and a Hamiltonian circuit, but each corner vertex is connected by edges to only two other vertices.

SOLVED PROBLEMS

Hamiltonian Paths and Circuits

1. For each of the graphs in Fig. 7-15, find a Hamiltonian circuit. If there is no Hamiltonian circuit, find a Hamiltonian path. If there is no Hamiltonian path, say so.

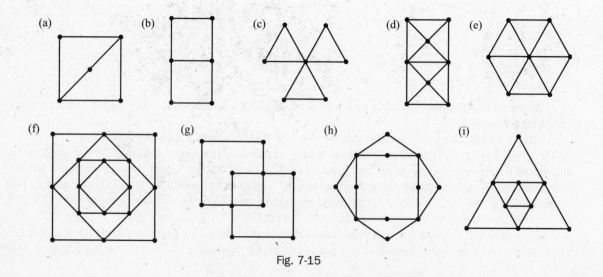

Fig. 7-15

2. Beat the Icosian game by finding a Hamiltonian circuit through all the vertices of the dodecahedron graph in Fig. 7-13(b) or 7-13(c).

Answers

1. Possible paths (there are many) are highlighted in Fig. 7-16.

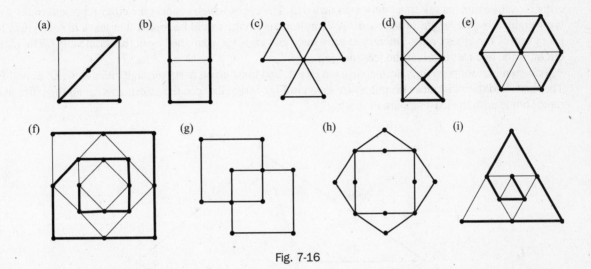

Fig. 7-16

(a) Hamiltonian path but no circuit
(b) Hamiltonian circuit
(c) No Hamiltonian path (thus no circuit either)
(d) Hamiltonian circuit
(e) Hamiltonian circuit
(f) Hamiltonian path, but no circuit
(g) No Hamiltonian path or circuit
(h) No Hamiltonian path or circuit
(i) Hamiltonian path, but no circuit

2. A solution is highlighted in Fig. 7-17.

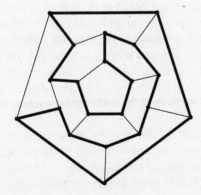 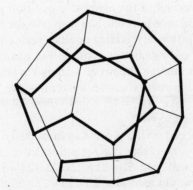

Fig. 7-17

The Traveling Salesman Problem

There are many situations where a graph clearly has a Hamiltonian circuit, but the goal is to find the best one. A *weighted graph* is one for which each edge is assigned a number (usually representing cost, distance, or time). An ideal Hamiltonian circuit is often one for which the number sum for all the edges used is minimal.

The classic version of this problem is called the *traveling salesman problem.* Each vertex represents a city or location that the salesman must visit on a trip. The edges which connect the cities represent roads and are weighted by the length of the roads. A Hamiltonian circuit would be a path that takes the salesman to every one of the cities on his itinerary exactly once and ends up where he began (his home city). The ideal circuit is the one that involves the least driving.

For example, suppose the salesman lives in city A, and must go on a trip through cities B, C, D, E, and F. The roads and their distances (in miles) are given in Fig. 7-18. How can the salesman visit all the cities and return home with the least amount of driving?

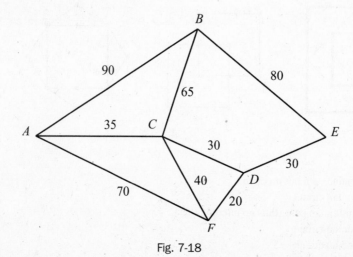

Fig. 7-18

If the salesman makes his circuit by going to cities *ACBEDFA* in that order, then he will have traveled a total distance of $35+65+80+30+20+70=300$ miles. If, instead, he travels *ABEDCFA*, then the total distance will be $90+80+30+30+40+70=340$ miles. If he travels *ABEDFCA*, then the distance will be $90+80+30+20+40+35=295$ miles. This is the best circuit so far, but is it *optimal*, the best possible?

The *brute force* technique to answer this question is to list out all the possible Hamiltonian circuits, calculate the total distance for each, and then take the lowest one. For small graphs, this can be a reasonable approach. Unfortunately, adding another vertex to the graph generally multiplies the difficulty of the problem by the old number of vertices. For example, suppose it takes you an hour to list out and compute all the Hamiltonian circuits on a complete graph with six vertices. Adding one more vertex will make the problem take about six times longer to solve. Increasing the graph again, to eight vertices, will make the problem take $6\times7=42$ hours; increasing to nine vertices will make the problem take $42\times8=336$ hours; and so on.

Computers can speed up the process tremendously, but they quickly run into the same problem. For example, if it takes a computer one second to solve any 20-vertex traveling salesman problem, then a 25-vertex problem will take $20\times21\times22\times23\times24=5,100,480$ seconds$=1,416.8$ hours ≈ 59 days. A 30-vertex problem would then take $25\times26\times27\times28\times29=14,250,600$ times longer, approximately 2.3 million years.

There are also *approximation techniques* that work to find a circuit that is close to optimal. These can obtain reliable answers in a reasonable amount of time, even on enormous graphs.

So far, however, no one has been able to develop a method for finding the optimal Hamiltonian circuit for a weighted graph that does not become several times more complicated with each additional vertex.

Mathematicians and computer scientists have found hundreds of equivalent problems. If a quick technique is found to solve any one of these problems, then the same technique will be able to solve all of the other problems. If no quick technique were ever possible, this would be useful to know. Many people now consider settling this matter one way or the other to be one of the most important unsolved problems in mathematics, known as the *P=NP* problem.

The bridges of Königsberg problem seemed impossibly hard until Euler found a new way to look at the situation. Could the *P=NP* problem also have an easy solution? Is there a clever way to represent the information from a traveling salesman problem which makes the optimal circuit stand out clearly? If so, then hundreds of important tasks (including delivering packages, connecting telephone calls, scheduling sporting competitions, and much more) will suddenly become easier and more efficient. This illustrates a beautiful aspect of mathematics: there is always the potential for a clever idea or new perspective to completely change the world.

SOLVED PROBLEMS

The Traveling Salesman Problem

1. For each of the weighted graphs in Fig. 7-19, name all of the Hamiltonian circuits that begin and end at *A*, and find the total weight of each.

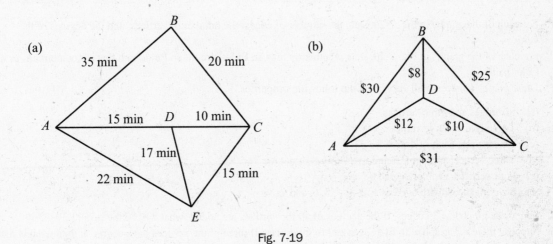

Fig. 7-19

Answers

1. (a) There are four Hamiltonian circuits that begin and end at *A*. The circuit formed by *ABCDEA* will take a total of $35+20+10+17+22=104$ minutes. The reverse of that circuit, *AEDCBA*, also has weight 104 minutes. The circuit formed by *ABCEDA* or its reverse, *ADECBA*, will have total weight $35+20+15+17+15=102$ minutes.

   (b) There are six Hamiltonian circuits: *ABCDA* and its reverse, *ADCBA* (total weight $77); *ABDCA* and its reverse, *ACDBA* (total weight $79); and *ADBCA* with its reverse, *ACBDA* (total weight $76).

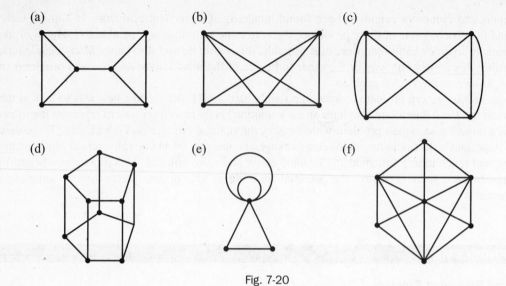

Fig. 7-20

SUPPLEMENTAL PROBLEMS

1. For each of the graphs in Fig. 7-20, state the number of edges, the number of vertices, and the degrees of the vertices.

2. For each of the graphs in Fig. 7-20, state whether there is an Euler circuit, an Euler path that is not a circuit, or no Euler path.

3. Draw graphs (answers will vary) with the following properties:

   (a) Six edges and three vertices
   (b) Two edges and five vertices
   (c) Six vertices, each of degree 3
   (d) Three vertices of degree 5 and five vertices of degree 3
   (e) Four vertices, each of degree 6
   (f) Seven vertices, with degrees 1, 2, 3, 4, 5, 6, and 7

4. Suppose a graph has 35 edges. If the degrees of all the vertices are added, what will be the result?

5. Suppose there is a graph with five edges and four vertices. If three of the vertices have degree 3, then what is the degree of the last vertex? Explain your answer.

6. What is the degree of each vertex in the complete graph K_n with (a) $n =$ seven vertices, (b) $n = 10$ vertices, (c) $n = 40$ vertices, and, in general, (d) n vertices?

7. Draw the complete bipartite graphs (a) $K_{5,2}$, (b) $K_{3,1}$, and (c) $K_{5,3}$.

8. What are the degrees of the vertices in the complete bipartite graphs (a) $K_{5,4}$, (b) $K_{3,7}$, (c) $K_{2,1}$, and, in general, (d) $K_{m,n}$?

9. For each of the graphs in Fig. 7-21, state if there is an Euler circuit, an Euler path that is not a circuit, or no Euler path. If there is an Euler path but not a circuit, name the vertices where every Euler path must begin and end.

10. Identify whether the graphs in Fig. 7-21 have a Hamiltonian circuit, a Hamiltonian path but no circuit, or else no Hamiltonian paths. If there is a Hamiltonian path or circuit, describe one (answers will vary) by naming the order in which the vertices are visited.

11. Is it possible to trace a pencil along each edge of a cube exactly once without lifting up the pencil? If it is possible, can you end up where you started? If it is not possible, explain why not.

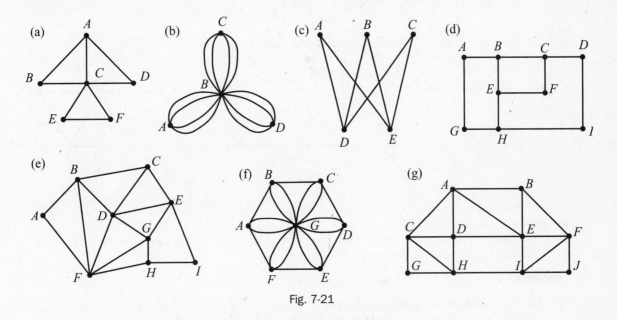

Fig. 7-21

12. Suppose you want to trace a pencil along every edge of a die and also draw an "X" across each face, as illustrated in Fig. 7-22. Can this be done without retracing or lifting up the pencil?

Fig. 7-22

13. Illustrate a Hamiltonian circuit using the edges of a cube.

14. A public works department wants to design a route for their sand truck to drive through the streets of the town during snowstorms. Ideally, the route would go over every street once and end up back at the city garage. In graph theory terms, what are they looking for?

15. There are 20 locations in a building that a night watchman must periodically check. It would be nice if there was a way for him to walk that went through each location only once. In graph theory terms, what would this watchman like?

16. A woman wants to drive from Florida to Maine in such a way that she goes through each of the 48 continental states. She would prefer to avoid driving through any state more than once. In graph theory terms, what does she want?

17. A puzzle called the *knight's tour* requires a knight from a chess set, all alone on a chessboard, to visit each square on the board exactly once before returning to where it started. In graph theory terms, what is needed to solve this puzzle? Describe how the graph might be constructed.

18. List all of the Hamiltonian circuits that begin at *A* for the weighted graphs in Fig. 7-23, and give the total weight of each.

19. Suppose a spy organization creates a weighted graph based on people's calling patterns. Every vertex represents a telephone. An edge between two vertices represents a phone call between the two telephones. The number attached to each edge is the number of calls between the two phones.

(a) 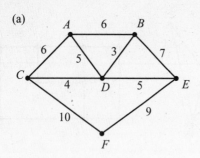   (b) 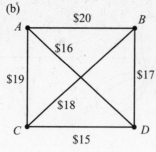   (c)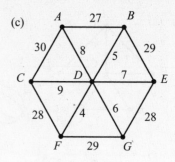

Fig. 7-23

(a) What would the degree of a vertex represent?

(b) If the numbers of all the edges attached to a phone were added, what would this represent?

(c) What would a path in this graph represent?

Answers

1.   (a)   Nine edges and six vertices, all of degree 3
      (b)   Nine edges and five vertices: three with degree 4 and two with degree 3
      (c)   Eight edges and four vertices, all of degree 4
      (d)   15 edges and ten vertices, all of degree 3
      (e)   Five edges and three vertices: two with degree 2 and one with degree 6
      (f)   15 edges and seven vertices: three with degree 3, three with degree 5, and one with degree 6

2.
      (a)   No Euler path
      (b)   An Euler path
      (c)   An Euler circuit
      (d)   No Euler path
      (e)   An Euler circuit
      (f)   No Euler path

3.   Possible answers are given in Fig. 7-24. Note that it is impossible for (b) to be a connected graph.

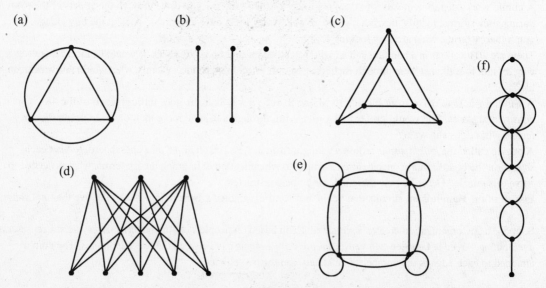

Fig. 7-24

4. 70

5. The last vertex must have degree 1 because the sum of all the degrees must be double the number of edges: ten.

6. (a) Six, (b) nine, (c) 39, and (d) $n-1$

7. Shown in Fig. 7-25.

(a)   (b)   (c)

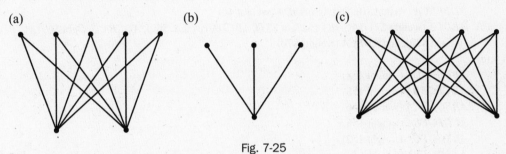

Fig. 7-25

8. (a) $K_{5,4}$ has five vertices of degree 4 and four vertices of degree 5.
   (b) $K_{3,7}$ has three vertices of degree 7 and seven vertices of degree 3.
   (c) $K_{2,1}$ has two vertices of degree 1 and one vertex of degree 2.
   (d) In general, $K_{m,n}$ has m vertices of degree n and n vertices of degree m.

9. (a) There are Euler paths beginning and ending with A and C.
   (b) There are Euler circuits.
   (c) There are Euler paths beginning and ending with D and E.
   (d) There are no Euler paths.
   (e) There are no Euler paths.
   (f) There are Euler circuits.
   (g) There are Euler paths beginning and ending with B and E.

10. Note that most of the graphs have several Hamiltonian paths or circuits, but only one example will be given.

    (a) Hamiltonian path: *BADCEF*
    (b) No Hamiltonian path
    (c) Hamiltonian path: *ADBEC*
    (d) Hamiltonian circuit: *ABEFCDIHGA*
    (e) Hamiltonian circuit: *ABCDEIHGFA*
    (f) Hamiltonian circuit: *ABCDEFGA*
    (g) Hamiltonian circuit: *ABFJIHGCDEA*

11. A way to trace all the edges exactly once without lifting up the pencil would be an Euler path. The graph of a cube's edges has eight vertices of odd degree, and thus this challenge is not possible.

12. This is possible because all of the vertices have degree 6 (even).

13. Shown in Fig. 7-26.

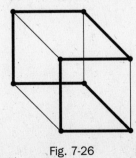

Fig. 7-26

14. An Euler circuit
15. A Hamiltonian circuit
16. A Hamiltonian path
17. This puzzle is looking for a Hamiltonian circuit. To construct the graph, draw 64 vertices (one for each square on the board), and connect two with an edge whenever it is possible for a knight to jump from one to the other.
18. (a) *ABDEFCA* (weight 39), *ACFEDBA* (weight 39), *ABEFCDA* (weight 41), *ADCFEBA* (weight 41), *ADBEFCA* (weight 40), and *ACFEBDA* (weight 40)

    (b) *ABDCA* (weight $71), *ACDBA* (weight $71), *ABCDA* (weight $69), *ADCBA* (weight $69), *ACBDA* (weight $70), and *ADBCA* (weight $70)

    (c)

    ABEGFCDA (weight 158)
    ADCFGEBA (weight 158)
    ABEGFDCA (weight 156)
    ACDFGEBA (weight 156)
    ABEGDFCA (weight 152)
    ACFDGEBA (weight 152)
    ABEDGFCA (weight 156)
    ACFGDEBA (weight 156)
    ABDEGFCA (weight 154)
    ACFGEDBA (weight 154)
    ADBEGFCA (weight 157)
    ACFGEBDA (weight 157)

19. (a) The degree of a vertex would represent the number of phones with which that telephone had had a phone call.

    (b) This would be the total number of phone calls (both incoming and outgoing) made with this phone.

    (c) A path would represent a list of telephone numbers where each phone on the list had been used in calls with the phones before and after it on the list. This could be used to identify a connection between two people who had never directly spoken on the phone.

CHAPTER 8

Financial Mathematics

Simple Interest

Banks make money by charging more for loans than they pay for savings. The money that is borrowed or saved is called the *principle*. The money that is charged or paid is called *interest*. The interest is calculated as a percentage of the principle, represented by the *interest rate*. The *annual percentage rate* (APR) is the rate for money borrowed or lent for one year.

Interest is *simple* when it only applies to the original principle invested or borrowed. This means that the amount of interest is the same each year. For example, $10,000 invested at 8% APR simple interest will earn $(0.08) \times 10,000 = \$800$ each year. After two years, the investment will be worth $10,000 + 2 \times 800 = \$11,600$. After $3\frac{1}{2}$ years, the investment will be worth $10,000 + (3.5) \times 800 = \$12,800$.

SOLVED PROBLEMS

Simple Interest

Compute the final value of each investment:

1. $5,000 invested for one year at 4% APR, simple interest
2. $800 invested for five years at 6% APR, simple interest
3. $24,000 invested for $8\frac{1}{2}$ years at 5% APR, simple interest

Answers

1. Each year the interest will be $(0.04) \times 5,000 = \$200$, so after one year the value will be $5,200.
2. Each year the interest will be $(0.06) \times 800 = \$48$, so the value after five years will be $800 + 5 \times 48 = \$1,040$.
3. Each year the interest will be $(0.05) \times 24,000 = \$1,200$, so the value after $8\frac{1}{2}$ years will be $24,000 + (8.5) \times 1,200 = \$34,200$.

Compound Interest

Most debts and investments earn *compound interest*, where interest is earned on both the original principle and the interest up to that point. An investment with compound interest will be worth more every year and thus will earn more interest.

The exact value of a compound investment depends upon how frequently the interest is added, usually yearly, quarterly, monthly, or daily. The rate of interest is the APR divided by the number of times the interest is calculated in a year. For example, an investment at 6% APR, compounded monthly, will earn $\frac{6\%}{12} = \frac{0.06}{12} = 0.005$ each month. Similarly, a 12% APR calculated daily results in $\frac{12\%}{365} = \frac{0.12}{365} = 0.0003288$ each day.

If

A = the amount initially invested (or borrowed),

n = the number of times interest is compounded each year,

$r = \dfrac{APR}{n}$ = the rate of interest for each pay period,

y = the number of years elapsed, and

F = the final value of the investment (or final amount of debt),

then $F = A(1+r)^{n \cdot y}$.

For example, suppose \$5,000 is put in a savings account at 6% APR, invested monthly, for ten years.

Here, $A=5{,}000$, $n=12$, $r = \dfrac{0.06}{12} = 0.005$, and $y=10$, so $F = 5{,}000 \times (1+0.005)^{12 \times 10} = 5{,}000(1.005)^{120} = 5{,}000 \times (1.819397) = \$9{,}096.98$.

The initial amount of money A will be negative if it represents a debt.

SOLVED PROBLEMS

Compound Interest

1. Compute the final value for each of the following:

    (a) \$6,000 invested at 8% APR, compounded monthly for five years
    (b) \$40,000 invested at 5% APR, compounded daily for seven years
    (c) A debt of \$8,000 at 12% APR, compounded monthly for two years

2. What will the final amount be for \$10,000 invested for five years at 6% APR, compounded (a) yearly, (b) quarterly, (c) monthly, and (d) daily?

    #### Answers

    1. (a) $A=6{,}000$, $n=12$ (monthly interest), $r = \dfrac{0.08}{12} \approx 0.00667$, and $y=5$, so $F = 6{,}000(1+0.00667)^{12 \times 5} = \$8{,}940.85$.

        (b) $A=40{,}000$, $n=365$, $r = \dfrac{0.05}{365} \approx 0.000137$, and $y=7$, so $F = 40{,}000(1+0.000137)^{365 \times 7} = \$56{,}763.33$.

        (c) $A=-8{,}000$ (debts are negative), $n=12$, $r = \dfrac{0.12}{12} = 0.01$, and $y=2$, so $F = -8{,}000(1+0.01)^{12 \times 2} = -\$10{,}157.88$.

    2. In each case, $A=10{,}000$ and $y=5$.

        (a) $n=1$ and $r=0.06$, so $F = 10{,}000(1+0.06)^{1 \times 5} = \$13{,}382.26$.

        (b) $n=4$ and $r = \dfrac{0.06}{4} = 0.015$, so $F = 10{,}000(1+0.015)^{4 \times 5} = \$13{,}468.55$.

        (c) $n=12$ and $r = \dfrac{0.06}{12} = 0.005$, so $F = 10{,}000(1+0.005)^{12 \times 5} = \$13{,}488.50$.

        (d) $n=365$ and $r = \dfrac{0.06}{365} \approx 0.00016438$, so $F = 10{,}000(1+0.00016438)^{365 \times 5} = \$13{,}498.17$.

Annual Percentage Yield

The *annual percentage yield* (APY) for an investment with compound interest is the rate of simple interest which earns the same amount in one year. For example, 5.25% APR compounded monthly will turn $100, in one year, into $100\left(1+\dfrac{0.0525}{12}\right)^{12 \times 1}$ = \$105.38. An investment with a simple interest rate of 5.38% would earn the same amount in a year. Thus, 5.25% APR compounded monthly has an APY of 5.38%.

The amount of the principle does not affect the calculation of APY, so it is most convenient to use $A = \$100$.

SOLVED PROBLEMS

Annual Percentage Yield

1. Compute the APY for each of the following:

    (a) 4% compounded daily
    (b) 17% compounded monthly
    (c) 4.5% compounded monthly

2. Which is a better investment: 5% APR compounded quarterly or 4.9% APR compounded daily?

    Answers

    1. (a) $100 becomes $100\left(1+\dfrac{0.04}{365}\right)^{365 \times 1}$ = \$104.08, so the APY is 4.08%.

        (b) $100 becomes $100\left(1+\dfrac{0.17}{12}\right)^{12 \times 1}$ = \$118.39, so the APY is 18.39%.

        (c) $100 becomes $100\left(1+\dfrac{0.045}{12}\right)^{12}$ = \$104.59, so the APY is 4.59%.

    2. Under the first investment, $100 becomes $100\left(1+\dfrac{0.05}{4}\right)^{4}$ = \$105.09. Under the second, $100 becomes $100\left(1+\dfrac{0.049}{365}\right)^{365}$ = \$105.02. The first has a higher APY and is thus a better investment.

Compound Interest with Payments

The compound interest formula only applies to situations where the principle is left untouched to accrue interest. A different formula is used when people add or subtract more money periodically. We will use the variable P to represent the *payment*, the amount added each period.

In a savings plan, both A and P are positive: money is put in the bank, and more is added periodically. A loan starts with a negative amount of money A (a debt), to which a positive P is paid periodically (to pay off the debt). An *annuity* is a positive amount A of money from which regular withdrawals are made (negative P). When both A and P are negative, the debt piles up. Such a situation occurs when more each month is charged than paid off on a credit card.

If the initial amount is A, the payments P are made n times a year for y years, and the periodic interest is $r = \dfrac{APR}{n}$, then the final amount F will be $F = \dfrac{(rA + P)(1 + r)^{n \cdot y} - P}{r}$.

For example, suppose a college fund is started with $500. This is put in a savings account at 4% APR, compounded monthly. Every month for 18 years, $100 more is added to the account. Here, $A = 500$, $n = 12$, $y = 18$, $r = \dfrac{0.04}{12} \approx 0.003333$, and $P = 100$. Thus, the amount in the account after the 18 years will be

$$F = \frac{(0.003333 \times 500 + 100)(1 + 0.003333)^{12 \times 18} - 100}{0.003333} = \frac{(101.6665)(2.0518) - 100}{0.003333} = \$32,583.06.$$

SOLVED PROBLEMS

Compound Interest with Payments

1. Suppose a person owes $2,000 on a credit card with 18% APR, compounded monthly. If the person stops using the card and pays off $50 each month, how much will still be owed after three years?
2. Suppose a person puts $1,000,000 in a bank account that pays 4% APR, compounded daily, and takes out $200 each day. How much will be left after ten years?
3. A person with $800 owed on a credit card decides to make no payments for a year. This makes the interest rate jump to 28% APR (compounded monthly). Each month, the account is penalized a $35 fee. If the person spends a total of $50 on the card each month, what will the balance be at the end of the year?
4. A smoker quits and puts the $5 previously spent each day on cigarettes in an account that earns 3% APR, compounded daily. How much will be in the account after four years?

Answers

1. $A = -2,000$ (a debt), $n = 12$ (monthly compounding), $y = 3$, $r = \dfrac{0.18}{12} = 0.015$, and $P = 50$ (positive because the debt is being paid off). Thus, $F = \dfrac{((0.015) \times (-2000) + 50)(1 + 0.015)^{12 \times 3} - 50}{0.015} = -\$1,054.48.$ The person will still owe over $1,000, even after a total of $36 \times 50 = \$1,800$ in payments.

2. $A = 1,000,000$ (positive savings), $n = 365$, $y = 10$, $r = \dfrac{0.04}{365} \approx 0.0001096$, and $P = -200$ (negative because money is subtracted every day). Thus, $F = \dfrac{((0.0001096) \times 1,000,000 + (-200))(1 + 0.0001096)^{365 \times 10} - (-200)}{0.0001096} =$
    $\dfrac{-90.4(1.4919) + 200}{0.0001096} = \$594,272.26.$ The account is down just over $400,000, even after subtracting $365 \times 10 \times 200 = \$730,000$, all because of interest.

3. $A = -800$ (a debt), $n = 12$, $y = 1$, $r = \dfrac{0.28}{12} \approx 0.02333$, and $P = -85$ (the balance goes down $35 for the fee and $50 for the spending each month). Thus, $F = \dfrac{((0.02333) \times (-800) + (-85))(1.02333)^{12} - (-85)}{0.02333} =$
    $-\$2,216.68.$ The person has spent $50 \times 12 = 600$, but the fees and interest have contributed $2,216.68 - 600 - 800 = \$816.68$ more.

4. $A = 0$ because the person starts with an empty bank account. The payment is $P = 5$, a positive amount deposited each day. The other variables are $y = 4$, $n = 365$, and $r = \dfrac{0.03}{365} \approx 0.0000822$. Thus
    $$F = \frac{((0.0000822) \times 0 + 5)(1.0000822)^{365 \times 4} - 5}{0.0000822} = \$7,755.76.$$ In instances like this, where the initial amount is $A = 0$, the formula is sometimes written $F = \dfrac{P \times \left((1 + r)^{n \times y} - 1 \right)}{r}.$

Saving for a Goal

Suppose you have no money and want to have F dollars at the end of y years. If you put money into a savings account n times a year and earn $r = \dfrac{APR}{n}$ periodic interest, then your regular payments will have to be

$$P = \frac{rF}{(1+r)^{n \cdot y} - 1}.$$

For example, suppose a 30-year-old person wants to have $1,000,000 in the bank upon retiring at 65 years old. If the best savings account available offers 5% APR, how much will this person have to deposit each month? Here, $n=12$, $y=65-30=35$, $F=1,000,000$, and $r = \dfrac{0.05}{12} \approx 0.0041667$, so the payments

will have to be $P = \dfrac{(0.0041667) \times 1,000,000}{(1.0041667)^{12 \times 35} - 1} = \$880.20$ each month. This might seem like too much,

but without interest the savings would only be $880 \times 12 \times 35 = \$396,600$.

SOLVED PROBLEMS

Saving for a Goal

1. How much would a person have to save each day in a 4% APR account in order to have $10,000 at the end of two years?
2. How much would parents have to put aside each month in a 6% APR money market fund in order to have $100,000 for their child's college education in 18 years?

    Answers

    1. $n=365$, $y=2$, $F=10,000$, and $r = \dfrac{0.04}{365} \approx 0.0001096$, so $P = \dfrac{(0.0001096) \times 10,000}{(1.0001096)^{365 \times 2} - 1} = \$13.16$ per day.

    2. $n=12$, $y=18$, $F=100,000$, and $r = \dfrac{0.06}{12} = 0.005$, so $P = \dfrac{(0.005) \times 100,000}{(1.005)^{12 \times 18} - 1} = \$258.16$ per month.

Paying Off a Loan

If a loan of A (negative) dollars is paid off with n payments a year for y years at $r = \dfrac{APR}{n}$ periodic interest, then the payments will have to be $P = \dfrac{rA}{(1+r)^{-n \cdot y} - 1}$.

For example, suppose a person wants to pay off a $12,000 car loan at 9% APR over four years with monthly payments. Here $A = -12,000$, $n=12$, $y=4$, and $r = \dfrac{0.09}{12} = 0.0075$. The monthly payments will have to

be $P = \dfrac{(0.0075) \times (-12,000)}{(1.0075)^{-12 \times 4} - 1} = \dfrac{-90}{0.6986 - 1} = \$298.61$.

SOLVED PROBLEMS

Paying Off a Loan

1. Suppose the APR for a 30-year fixed-rate mortgage is 6.5%. What would the monthly payments be for a house that costs (a) $200,000 or (b) $350,000?

Answers

1. In each case, $n = 12$, $y = 30$, and $r = \dfrac{0.065}{12} \approx 0.0054167$.

(a) $A = -200,000$ (a debt), so $P = \dfrac{(0.0054167) \times (-200,000)}{(1.0054167)^{-12 \times 30} - 1} = \$1,264.14$ per month.

(b) $A = -350,000$, so $P = \dfrac{(0.0054167) \times (-350,000)}{(1.0054167)^{-12 \times 30} - 1} = \$2,212.24$ per month.

The Time Required to Pay Off a Debt

In order to pay off a debt of A (negative) dollars by paying P (positive) dollars n times a year with an $r = \dfrac{APR}{n}$ periodic interest rate, it is necessary that $P > -rA$. Otherwise, the regular payments will not even cover the interest and the debt will grow. If the payments P are greater than the initial interest $-rA$, then

the number of years y required to pay off the debt will be $y = \dfrac{\log\left(\dfrac{P}{rA + P}\right)}{n \times \log(1 + r)}$.

For example, how long will it take a person to pay off \$1,200 of credit card debt at 18% APR by making the minimum payment of \$50 each month (and not using the card anymore)? Here, $A = -1,200$, $P = 50$,

$n = 12$, and $r = \dfrac{0.18}{12} = 0.015$, so $y = \dfrac{\log\left(\dfrac{50}{(0.015) \times (-1200) + 50}\right)}{12 \times \log(1.015)} = \dfrac{\log(1.5625)}{12 \times (0.006466)} \approx 2.5 \text{ years}.$

SOLVED PROBLEMS

The Time Required to Pay Off a Debt

1. The monthly payments on a 30-year, \$200,000 mortgage at 6.5% APR are officially \$1,264.14. If the person pays \$1,300 each month instead, how long will it take to pay off the debt?
2. A person agrees to pay off a \$10,000 debt by paying \$500 four times a year. If the interest rate is 8% APR, compounded quarterly, how long will it take to pay off the debt?

Answers

1. $A = -200,000$, $n = 12$, $P = 1,300$, and $r = \dfrac{0.065}{12} \approx 0.0054167$, so

$y = \dfrac{\log\left(\dfrac{1,300}{(0.0054167) \times (-200,000) + 1,300}\right)}{12 \times \log(1 + 0.0054167)} = \dfrac{\log(6)}{12\log(1.0054167)} = 27.64 \approx 27 \text{ years, } 8 \text{ months.}$

Thus, the extra \$36 each month will eliminate the last 28 months of payments (a total of $28 \times 1,264 = \$35,392$).

2. $A = -10,000$, $n = 4$, $P = 500$, and $r = \dfrac{0.08}{4} = 0.02$, so $y = \dfrac{\log\left(\dfrac{500}{(0.02) \times (-10,000) + 500}\right)}{4 \times \log(1 + 0.02)} = 6.449$

years ≈ 6 years, 6 months.

SUPPLEMENTAL PROBLEMS

1. Suppose an investment of $12,000 earns 8% simple interest. How much will it be worth after one year?
2. How much will $750 earn in six years at 5% simple interest?
3. What will the final value of $8,000 be after $12\frac{1}{2}$ years of 4% simple interest?
4. How much will $5,000 become in three years at 6% interest, compounded monthly?
5. What will the final value of $400 be when compounded daily at 8% interest for ten years?
6. In 1917, a $100 bond was purchased, earning 4% interest compounded quarterly. How much would it be worth in 2007?
7. What would happen to a single penny invested at 5% interest, compounded yearly, 500 years ago?
8. A student takes out a $2,000 loan at 3.1% interest, compounded monthly, and then leaves town. The loan company catches up with the student five years later. How much will the student owe then?
9. What is the annual percentage yield for (a) 6% compounded monthly, (b) 5% compounded daily, (c) 4.29% compounded monthly, and (d) 4.33% compounded quarterly?
10. Which is a better investment: 3.2% compounded daily or 3.25% compounded monthly?
11. Amy has $85 in a bank account with 3.5% APR. If she puts in $10 every day, how much will she have in five years?
12. Helen has an annuity of $200,000 with a 5% APR. Each month she takes out $1,800. How much will be left after ten years?
13. Michael takes out a $100,000 loan at 6% APR. If he pays $700 every month for ten years, how much will he still owe?
14. Brent has no money, but saves $100 each month in an account that earns 5.5% interest. How much will he have in the account after six years?
15. A credit card debt of $3,500 is ignored for two years, earning a $39 penalty each month, plus 32% interest. What will the end balance be?
16. A student could pay for an $8,000 semester of college with either a 3.25% APR student loan (compounded quarterly) or a 12% APR credit card (compounded monthly). What will the difference be if each is left to accrue interest for four years?
17. A 20-year, $200,000 mortgage at 6% APR demands monthly payments of $1,432.86. After the first payment is made, how much debt remains?
18. How much would you have to save each month in a 4% APR account in order to have $100,000 at the end of ten years?
19. A 16 year old wants to buy a $2,000 car in one year. How much must she save each day in a 5% APR savings account to reach this goal?
20. Maya has 40 years left before she retires. She would like to have $500,000 at that time. How much will she need to save each week in a 4% APR savings account?
21. What are the monthly payments on a $300,000 mortgage at 6.2% APR if paid over (a) 20 years, (b) 25 years, and (c) 30 years?
22. Becky wants to buy an $18,000 car. What will the monthly payments be if she pays 8% interest over five years?
23. The Klein family has a total of $9,000 in credit card debt. They transfer the balance to a 12% APR card and plan to pay it all off in four years. How much will they have to pay each month?
24. How long will it take to pay off $5,000 in debt at 14% interest by paying $75 each month?
25. The payments on a five-year, $12,000 car loan at 9% APR are officially $249.10. If Jared pays $275 each month instead, how long will it take to pay off the debt?
26. Emily owes $2,200 on a credit card with a 12% APR. The minimum payment is $25 each month. How long will it take to pay off the debt by making the minimum payment each month?
27. Sarah owes her father $1,000. She pays him $10 each week. He charges her 2% interest. How long will it take to pay off the debt?

Answers

1. $12,960
2. $225 of interest
3. $12,000
4. $5,983.40
5. $890.14
6. $3,594.96
7. It would become $393,232,618.
8. $2,334.85
9. (a) 6.17% APY, (b) 5.13% APY, (c) 4.38% APY, and (d) 4.4% APY
10. 3.2% compounded daily has an APY of 3.25%, while 3.25% compounded monthly has an APY of 3.3%, so the latter is better.
11. $20,044.46
12. $49,893.80
13. $67,224.13
14. $8,507.34
15. −$7,870.21
16. The student loan debt would be $9,105.84, while the credit card debt would be $12,897.81: a difference of $3,791.97.
17. $199,567.14
18. $679.12
19. $5.34
20. $97.37
21. (a) $2,184.05, (b) $1,969.75, and (c) $1,837.41
22. $364.98
23. $237
24. 10.8 years ≈ 10 years, 10 months
25. 4.42 years ≈ 4 years, 5 months
26. 17.76 years ≈ 17 years, 9 months
27. 1.96 years ≈ 2 years

CHAPTER 9

Probability

A *probability* is a number between 0 and 1 that indicates the likelihood of something happening. Events with high probability are more likely than those with low probability. Events that cannot happen have probability 0 and events that always happen have probability 1. For example, every day the sun rises with probability 1 and crashes into the earth with probability 0. Calculating probabilities can be useful in making good decisions.

Expectations

A flipped coin will land heads with probability $\frac{1}{2}$ and tails with probability $\frac{1}{2}$ (ignoring warped and trick coins). This means that heads is as likely to land up as tails. This does not mean that a coin flipped twice will land tails exactly once. Probabilities do not guarantee anything, but help to estimate long-term behavior. If the coin is flipped 1,000 times, for example, then we expect it to land heads approximately $\frac{1}{2} \cdot 1,000 = 500$ times.

If the coin lands heads 514 times, we will not be surprised, but if it lands heads 200 times, we will wonder if there is something odd about the coin.

In general, if something with probability p is attempted a large number N times, we will expect it to happen approximately $p \times N$ times.

As an example, suppose that 1,000 people in a city of 1,000,000 people have a terrible disease. Furthermore, suppose there is a test for this disease which is 99% accurate. Ignoring the costs and difficulties, why would it be a bad idea to test every single person in the city for the disease? When the test is run on the 1,000 people with the disease, we expect approximately $0.99 \times 1,000 = 990$ to get an accurate test result: that they have the disease. This means that we expect roughly ten with the disease to get inaccurate results stating that they do not have the disease: *false negatives*. When the 999,000 people without the disease take the test, we expect that about $0.99 \times 999,000 = 989,010$ will receive the truth: that they do not have the disease. The remaining $999,000 - 989,010 = 9,990$ will receive *false positive* results: that they have the disease when they do not. This means that we expect around $990 + 9,990 = 10,980$ people to test positive. Of these, roughly 9,990 will be false positives. Curiously, even though the test is 99% accurate, over 90% of the people who test positive will not have the disease. It is for this reason that tests are only given to people with a good reason to suspect that they might have a disease. On the other hand, there will be approximately 989,020 people who will test negative, ten of them receiving false negatives. This means that roughly $\frac{989,010}{989,020} \approx 99.999\%$ of the people who test negative will not have the disease. Thus, people who test negative can be fairly sure they do not have the disease but people who test positive should take the test again!

SOLVED PROBLEMS

Expectations

1. Suppose the probability of a rainy day is 5% in a certain town. How many days of rain should the townspeople expect each year?
2. Suppose a state lottery gives out 20 cents in prizes for every dollar received. If a person were to spend $10 on lottery tickets each week for five years, how much would he or she expect to win in prizes?
3. Suppose a company drug tests all 10,000 of its employees with a 98% accurate test. How many innocent people do you expect to fail the test if there are (a) 100, (b) 500, or (c) 1,000 drug users working for the company?

Answers

1. With 365 days in a year, the people should expect $0.05 \times 365 \approx 18$ days of rain each year.
2. Spending $10 per week for five years comes out to $10 \times 52 \times 5 = \$2,600$. The overall payout is 20 cents per dollar, so this person can expect to win approximately $0.2 \times 2,600 = \$520$ over the course of the five years.
3. (a) If there are 100 drug users at the company, we expect $0.98 \times 100 = 98$ of them to fail the test. Of the 9,900 nonusers, we expect $0.98 \times 9,900 = 9,702$ to pass and the remaining $9,900 - 9,702 = 198$ to get false positive results. This means that approximately 198 out of the $98 + 198 = 296$ people who fail will actually be innocent (about 67%).

   (b) If there are 500 drug users, about $0.98 \times 500 = 490$ will test positive. About 2% of the 9,500 nonusers will get false positives: $0.02 \times 9,500 = 190$. This means that roughly $\frac{190}{490 + 190} \approx 28\%$ of those who fail will be innocent.

   (c) If 1,000 employees take drugs, approximately $0.98 \times 1,000 = 980$ of them will test positive. Of the remaining 9,000 employees, about $0.02 \times 9,000 = 180$ will get a false positive result. Thus, about $\frac{180}{1160} \approx 16\%$ of those who fail the drug test will not be drug users.

Equally Likely Events

The easiest probabilities to calculate involve events that are equally likely to occur. For example, a flipped coin is as likely to be heads as tails. Similarly, if a die is rolled, then the six numbers are all equally likely to land up. When a card is drawn blindly from a deck, we suppose that each of the 52 cards has an equal likelihood of being drawn. In these cases, we suppose the die and coin are *fair*, weighted evenly so that all outcomes are equally likely, and that the person drawing the card knows nothing about the order of the deck.

When an action has N equally likely outcomes and we want any of M of them to happen, then our probability of success is $\frac{M}{N}$.

For example, what is the probability of rolling a prime number on a die? There are six equally likely numbers, and three of them are prime (2, 3, and 5). Thus, the probability of rolling a prime number is $\frac{3}{6} = \frac{1}{2}$.

People often make mistakes by assuming that events are equally likely. It is true that the sun will either rise or not each day, but these are not equally likely; the sun rises 100% of the time, not 50%. Casinos, for example, calculate the payouts on roulette bets by assuming that the ball is equally likely to land on any of the 37 (or 38) different numbers. In 1873, however, Monte Carlo lost $325,000 to an engineer named Joseph Jaggers who had carefully observed that nine of the numbers on a particular roulette wheel were turning up much more often than the others.

SOLVED PROBLEMS

Equally Likely Events

1.   What is the probability that a person's birthday will occur on a weekend?
2.   Suppose a kindergarten class has nine boys and 14 girls. What is the probability that a child chosen at random is a boy?
3.   Which is more likely to happen: rolling a 6 on a die or drawing a card from a deck and getting a king or a queen?
4.   A person playing poker holds 6♠, 7♣, 8♣, 9♥, and J♦. What is the probability of her completing a straight by discarding the jack and drawing either a 5 or a 10?

Answers

1.   We suppose that the birthday is equally likely to land on any of the seven days of the week. Because two of these are weekend days, the probability that a birthday falls on a weekend is $\frac{2}{7} \approx 29\%$.

2.   If the choice is truly random, then each of the 23 kids will have an equal chance of being chosen. The probability that one of the nine boys will be chosen is thus $\frac{9}{23} \approx 39\%$.

3.   The probability of rolling a 6 is $\frac{1}{6} \approx 17\%$ because there is one 6 and six equally likely outcomes. There are four kings and four queens in a deck of 52, so the probability of drawing one out is $\frac{8}{52} \approx 15\%$. Thus, it is more likely to roll a 6 than to draw a king or queen.

4.   The card that she draws is equally likely to be any of the 47 cards that she does not have in her hand (she is certain not to draw another 6♠, for example). There are eight cards that will enable her to complete the straight: four 5's and four 10's. Thus, her probability of success is $\frac{8}{47} \approx 17\%$.

Independent Events

Two different events are *independent* if knowing the outcome of one gives no clue about the outcome of the other. For example, suppose a coin and a die are dropped on a table. Looking at the coin gives us no information about the die and vice versa. These are thus independent events. On the other hand, a person becoming hungry and a person becoming thirsty are not independent events. People who are hungry are likely to be thirsty as well; thus, knowing one gives clues about the other.

When two events are independent, the probability of both happening is the product of their individual probabilities. For example, what is the probability of dropping a coin and a die on a table and getting both tails and a 1? The probability of the coin landing tails is $\frac{1}{2}$. The probability of the die rolling 1 is $\frac{1}{6}$. Because the coin and the die are independent, the probability of getting tails and a 1 at the same time is $\frac{1}{2} \cdot \frac{1}{6} = \frac{1}{12} = 8\frac{1}{3}\%$.

When several events are all independent, the probability of all happening is the product of their probabilities. For example, what is the probability of dropping four coins and having all land heads up? Each coin has an independent probability of $\frac{1}{2}$ to land heads. Thus, the probability that all will land heads is $\frac{1}{2} \cdot \frac{1}{2} \cdot \frac{1}{2} \cdot \frac{1}{2} = \left(\frac{1}{2}\right)^4 = \frac{1}{16} = 6.25\%$.

Independent Events

1. What is the probability of rolling two dice and getting two 6's?
2. What is more likely: rolling three 5's in a row with a die or flipping a coin heads seven times in a row?
3. What is the probability of drawing an ace out of a deck and then rolling a 1 on a die?
4. If a card is drawn randomly from a deck, what is the probability that the card is both red and a face card (jack, queen, or king)?
5. If a card is drawn randomly from a deck, what is the probability that the card is both red and a spade?
6. Suppose there is a 1% chance of seeing a shooting star in the sky. If shooting stars last 1 second in the sky, how long would you have to wait to see two shooting stars at the same time?

Answers

1. Each die has an independent probability of $\frac{1}{6}$ to roll 6. Thus, the probability that both roll 6 is

$$\frac{1}{6} \cdot \frac{1}{6} = \frac{1}{36} = 2\frac{7}{9}\%.$$

2. The probability of rolling one 5 is $\frac{1}{6}$. The probability of rolling three in a row is thus $\frac{1}{6} \cdot \frac{1}{6} \cdot \frac{1}{6} = \frac{1}{125}$.

   The probability of flipping a coin head seven times in a row is $\frac{1}{2} \cdot \frac{1}{2} \cdot \frac{1}{2} \cdot \frac{1}{2} \cdot \frac{1}{2} \cdot \frac{1}{2} \cdot \frac{1}{2} = \frac{1}{128}$. The second probability is higher, so flipping 7 heads is more likely.

3. These are independent events. The probability of drawing one of the four aces is $\frac{4}{52} = \frac{1}{13}$. The probability of rolling a 1 is $\frac{1}{6}$. Thus, the probability of doing both is $\frac{1}{13} \cdot \frac{1}{6} = \frac{1}{78} \approx 1.3\%$.

4. Half of the cards are red, so the probability of drawing a red card is $\frac{1}{2}$. There are 13 different ranks of cards, so the probability of drawing a jack, queen, or king is $\frac{3}{13}$. These events are independent because knowing the color of a card gives no information about its rank and vice versa. Thus, the probability of drawing a red face card is $\frac{1}{2} \cdot \frac{3}{13} = \frac{3}{26} \approx 12\%$. This could also have been calculated by counting the six red face cards in the deck of 52: $\frac{6}{52} = \frac{3}{26}$.

5. Drawing a red card and a spade are not independent events. If you know that a card is red, then it must be a heart or a diamond; there is no chance of it being a spade. Thus, there is a zero probability that a random card is both red and a spade.

6. Unless some phenomenon is causing a meteor shower, we can suppose that shooting stars are independent. Thus, the chance of two shooting stars occurring at the same time ought to be $\frac{1}{100} \cdot \frac{1}{100} = \frac{1}{10,000}$. Should you watch the sky for 10,000 seconds ($2\frac{7}{9}$ hours), you would expect to see this happen approximately once.

Complementary Events

The *complement* of an event is the collection of all the ways that it might not happen. For example, the complement of rolling a 6 is rolling a 1, 2, 3, 4, or 5. The complement of winning the lottery is not winning the lottery. The probability is 1 that either an event or its complement will occur. Thus, if an event has probability p, its complement has probability $1-p$.

There are many situations where calculating the complement of an event is easier than calculating the event itself. For example, what is the probability of rolling four dice and getting at least one 6? We will succeed if we roll one, two, three, or four 6's. Rather than calculate all these possibilities, it is much easier to calculate the complement: the probability that we do not roll a 6. In order to not roll a 6, we will need to roll 1–5 on each of four dice. The probability of rolling 1–5 on a die is $\frac{5}{6}$. Because the dice are independent, the probability of rolling a 1–5 on each of the four dice is $\frac{5}{6} \cdot \frac{5}{6} \cdot \frac{5}{6} \cdot \frac{5}{6} = \left(\frac{5}{6}\right)^4 = \frac{625}{1296} \approx 48.2\%$. Because there is a 48.2% chance of not rolling any 6's, there is a $100\% - 48.2\% = 51.8\%$ chance of rolling at least one 6.

SOLVED PROBLEMS

Complementary Events

1. What is the probability of rolling six dice and getting at least one 6?
2. What is the probability of flipping a coin three times and getting at least one head?
3. A person playing poker holds 2♥, 6♥, 10♣, 10♦, and Q♠. He decides to keep the two 10's and discard the rest. When he draws three more cards, what is the probability of getting at least another 10?
4. It is hard to tell if a holly plant is male or female, but you need to have both in a garden to get festive holly berries. Suppose that half of all holly plants are female. If you randomly buy (a) two, (b) three, (c) four, (d) five, or (e) six holly plants for your garden, what is the probability that you will get holly berries?

Answers

1. The probability of rolling a die and not getting a 6 is $\frac{5}{6}$. The probability of rolling six dice and not getting a single 6 is $\left(\frac{5}{6}\right)^6 = \frac{15,625}{46,656} \approx 33.5\%$. The probability of the complement, getting at least one 6, is thus $100 - 33.5 = 66.5\%$.

2. The complement of getting at least one head is getting all tails. The probability of getting all tails on three coins is $\left(\frac{1}{2}\right)^3 = \frac{1}{8}$. The probability of getting at least one head is thus $1 - \frac{1}{8} = \frac{7}{8} = 87.5\%$.

3. It is easiest to calculate the probability that this person draws three cards and gets no 10's. The probability that the first card drawn is not a 10 is $\frac{45}{47}$ because there are 47 cards that he has not seen yet and 45 are not 10's. The probability that the second card drawn is also not a 10 is $\frac{44}{46}$ because there will still be two 10's left in among the 46 remaining cards. The probability that the third card is also not a 10 will be $\frac{43}{45}$. Thus, the probability that no more 10's are drawn is $\frac{45}{47} \cdot \frac{44}{46} \cdot \frac{43}{45} \approx 87.5\%$. The probability that he will draw one or two more 10's is the complement: $100 - 87.5 = 12.5\%$.

4. There are two ways that a garden can fail to have both male and female holly plants: all can be male, or all can be female.

    (a) With two plants, the probability that both are male is $\frac{1}{2} \cdot \frac{1}{2} = \frac{1}{4}$, and the probability that both are female is the same: $\frac{1}{2} \cdot \frac{1}{2} = \frac{1}{4}$. Together, there is a $\frac{1}{4} + \frac{1}{4} = \frac{1}{2}$ chance of the plants being the same sex.

    The probability that the plants will be of opposite sex is thus $1 - \frac{1}{2} = \frac{1}{2} = 50\%$.

(b) With three plants, there is a $\left(\dfrac{1}{2}\right)^3 = \dfrac{1}{8}$ chance of the plants being all male and a $\dfrac{1}{8}$ chance of them being all female. This means that there is a $\dfrac{1}{8} + \dfrac{1}{8} = \dfrac{1}{4}$ chance that the garden will not have holly berries. The probability that there will be holly berries is $1 - \dfrac{1}{4} = \dfrac{3}{4} = 75\%$.

(c) With four plants, the probability that the plants are all male or all female is $\left(\dfrac{1}{2}\right)^4 + \left(\dfrac{1}{2}\right)^4 = \dfrac{1}{16} + \dfrac{1}{16} = \dfrac{1}{8}$, so the probability of getting holly berries is $1 - \dfrac{1}{8} = \dfrac{7}{8} = 87.5\%$.

(d) The probability that five holly plants are all the same sex will be $\dfrac{1}{32} + \dfrac{1}{32} = \dfrac{1}{16}$, so the chance of having at least one male and one female is $\dfrac{15}{16} = 93.75\%$.

(e) The probability of getting holly berries with six plants will be $1 - \dfrac{1}{32} = \dfrac{31}{32} \approx 97\%$.

Combinations and Permutations

One key to calculating probabilities is *combinatorics*: the study of counting.

A *permutation* is a way of putting different objects in order. For example, the permutations of the letters A, B, and C are ABC, ACB, BAC, BCA, CAB, and CBA. Here, there are three different letters, and so there are three different items that can go first. Once a first item has been chosen, there are two choices for what to put second. When the first and second items in line have been chosen, there is no more choice: the only remaining object must go last. The number of permutations of three objects is thus $3 \cdot 2 \cdot 1 = 6$.

Similarly, the number of permutations of five objects is $5 \cdot 4 \cdot 3 \cdot 2 \cdot 1 = 120$. This is called a *factorial* and abbreviated with an exclamation point, as in $5! = 5 \cdot 4 \cdot 3 \cdot 2 \cdot 1 = 120$. In general, the number of permutations of n objects is $n! = n \cdot (n-1) \cdot (n-2) \ldots 3 \cdot 2 \cdot 1$. Scientific calculators have an $n!$ or an $x!$ button for computing permutations. For example, the number of ways your top eight friends can be arranged in order is $8! = 40,320$.

If only some of the items are to be put in order, then you only multiply as many numbers as choices. For example, how many ways can first, second, and third prizes be awarded in a dog show with 25 dogs? There are 25 choices for first place, 24 for second, and 23 for third, for a total of $25 \cdot 24 \cdot 23 = 13,800$ different possibilities. This could be written as $\dfrac{25 \cdot 24 \cdot 23 \cdot 22 \cdot 21 \ldots 4 \cdot 3 \cdot 2 \cdot 1}{22 \cdot 21 \ldots 4 \cdot 3 \cdot 2 \cdot 1} = \dfrac{25!}{22!}$.

In general, if there are n different items and k of them are to be selected and set in order, there are $\dfrac{n!}{(n-k)!}$ ways to do this. This is often abbreviated as $_nP_k = P_{n,k} = \dfrac{n!}{(n-k)!}$. For example, if you want to form a top-eight list from 30 friends, there are $P_{30,8} = \dfrac{30!}{(30-8)!} = \dfrac{30!}{22!} \approx 2.3 \times 10^{11} = 230,000,000,000$ ways to do this.

There are many times when the order in which things are chosen does not matter. For example, if you invite five friends over for dinner, you do not need to say who was your first choice and who was your fifth choice. A *combination* is a way to choose things without an order. The number of combinations possible is found by taking the number of permutations and dividing by the number of ways the selected items can be rearranged. For example, how many ways can you choose five of eight friends to invite to a dinner? The number of ways they can be chosen in order is $P_{8,5} = \dfrac{8!}{(8-5)!} = \dfrac{8!}{3!} = 6,720$. The number of ways that the five friends can be rearranged is $5! = 120$. This means that each way of choosing the five friends has been counted 120 times, one for each possible order. Thus, the number of ways the friends can be chosen is $\frac{6,720}{120} = 56$.

The number of combinations which select k things from a total of n is $\frac{P_{n,k}}{k!} = \frac{n!}{(n-k)! \cdot k!}$. There are three

common ways to abbreviate this: $_nC_k = C_{n,k} = \binom{n}{k} = \frac{n!}{k! \cdot (n-k)!}$. For example, how many ways can three

of nine spices be selected for a soup? This is a combination if it does not matter what order the spices are

added to the soup. Thus, there are $\binom{9}{3} = \frac{9!}{3!(9-3)!} = \frac{9!}{3! \cdot 6!} = \frac{362,880}{6 \cdot 720} = 84$ different ways to choose three
of the nine spices.

Many calculators have both $_nC_k$ and $_nP_k$ buttons for calculating permutations and combinations.

SOLVED PROBLEMS

Permutations and Combinations

1. A wedding photographer wants to take a photo with the six members of the wedding party lined up in every possible way. How many photos must be taken?
2. How many ways can first, second, and third places be awarded to the 52 contestants in a beauty contest?
3. A child is allowed to invite three friends to the movies. If she has ten friends at school, how many different ways can she choose three to accompany her?
4. How many different five-card poker hands are there?

Answers

1. Because order matters, this is a permutation. There are $6! = 720$ ways to arrange six people in a line. Thus, 720 photos must be taken.

2. This is a permutation because the order matters; thus, there are

    $$_{52}P_3 = \frac{52!}{(52-3)!} = \frac{52!}{49!} = 52 \cdot 51 \cdot 50 = 132,600 \text{ ways the three winners can be selected.}$$

3. This is a combination problem because the order does not matter. The girl has

    $$_{10}C_3 = \frac{10!}{3! \cdot (10-3)!} = \frac{10!}{3! \cdot 7!} = \frac{3628800}{6 \cdot 5,040} = 120 \text{ ways she can choose three of her ten friends.}$$

4. The order of the cards does not matter, so this is a combination. The number of ways to choose five of the 52 cards in a deck is $_{52}C_5 = \frac{52!}{5!47!} = 2,598,960$.

Probabilities with Combinations

We can now use combinatorics to solve a number of probability problems.

Suppose a classroom has 12 girls and nine boys in it, but the teacher seems to always call on boys. If four different students are randomly called upon, what is the probability that all of them are boys? There are 21

students in the class, so there are $_{21}C_4 = \frac{21!}{4!17!} = 5,985$ ways that four could be chosen. The number of

ways four of the nine boys could be chosen is $_9C_4 = \frac{9!}{4!5!} = 126$. If everything is random and all of the combinations are equally likely, then the probability of choosing four boys is only $\frac{126}{5,985} \approx 2.1\%$. The teacher is obviously not choosing the four students at random.

What is the probability of being dealt a flush (all cards of the same suit) in five-card draw poker? The total number of five-card hands is $_{52}C_5 = 2,598,960$. The number of ways to choose a five-card hand of all diamonds is $_{13}C_5 = 1,287$ because there are 13 diamonds in the deck. There is the same number of all-hearts, all-clubs, and all-spades hands, for a total of $4 \times 1,287 = 5,148$ different way to get a flush. The probability of being dealt a flush is thus $\frac{5,148}{2,598,960} \approx 0.198\%$.

What is the probability of being dealt a three-of-a-kind? To be dealt three aces, you need three of the four aces, which can be picked in $_4C_3 = 4$ ways. The remaining two cards can be chosen from any of the 48 non-aces in $_{48}C_2 = 1,128$ ways. Any combination of these will result in a hand with three aces; thus, there are $4 \times 1,128 = 4,512$ different hands with three aces. The number of hands with three kings, three queens, or three cards of any one rank is also the same, for a grand total of $13 \times 4,512 = 58,656$ different three-of-a-kind hands. Because there are 2,598,960 different five-card hands, the probability of being dealt a three-of-a-kind is $\frac{58,656}{2,598,960} \approx 2.26\%$.

It is because of their probabilities that a flush beats a three-of-a-kind in poker. In general, the kind of hand with the lower probability is the stronger hand and wins the game.

SOLVED PROBLEMS

Probabilities with Combinations

1. Suppose ten coins are dropped on a table. What is the probability that exactly four land heads?
2. If a coin is flipped 100 times, you will expect that heads will come up approximately 50 times. What, however, is the probability of getting exactly 50 heads?
3. What is the probability of being dealt a four-of-a-kind in five-card draw poker?
4. Suppose grenades are shipped in crates of 100. To test the quality of a crate, three grenades are randomly taken out and tested. If any of the three fail to detonate, the crate is rejected. Suppose a crate has five duds. What is the probability that it will pass the test?
5. Suppose a drawer contains ten black socks, eight white socks, and two red socks. If two socks are drawn out randomly, what is the probability that they match?

Answers

1. There are two different ways each coin can land, for a total of $2^{10} = 1,024$ possible outcomes. The number of ways that four coins land heads is the number of ways four of the ten coins can be chosen:
    $$_{10}C_4 = \frac{10!}{4! \cdot 6!} = 210.$$ Thus, the probability that exactly four coins land heads up is $\frac{210}{1,024} \approx 20.5\%$.
2. There are $2^{100} \approx 1.27 \times 10^{30}$ possible outcomes of the 100 coin flips. The number of ways 50 of the 100 can be heads is $_{100}C_{50} \approx 10^{29}$. Thus, the probability that exactly 50 of the coin flips will be heads is approximately
    $$\frac{10^{29}}{1.27 \times 10^{30}} = \frac{1}{1.27 \times 10} \approx 7.9\%.$$
3. There is only one way to choose all four aces from a deck. The fifth card, however, can be any one of the 48 remaining cards. Thus, there are 48 different hands with four aces. Because there are 13 different ranks of cards, there are a total of $13 \times 48 = 624$ different four-of-a-kind hands. The probability of being dealt one out of the 2,598,960 possible hands is $\frac{624}{2,598,960} = \frac{1}{4,165} \approx 0.024\%$.
4. The number of ways three of the 100 grenades can be randomly chosen is $_{100}C_3 = 161,700$. The number of ways these three grenades can be chosen from the 95 good grenades is $_{95}C_3 = 138,415$. Thus, the probability that the three grenades tested will all detonate is $\frac{138,415}{161,700} \approx 85.6\%$.

5. There are $_{10}C_2 = 45$ ways to pick two of the ten black socks, $_8C_2 = 28$ ways to pick two of the eight white socks, and $_2C_2 =$ one way to pick the two red socks. In general, there are $_{20}C_2 = 190$ ways to pick two of the 20 total socks. Thus, the probability of drawing out a matching pair is $\dfrac{45 + 28 + 1}{190} = \dfrac{74}{190} \approx 38.9\%$.

SUPPLEMENTAL PROBLEMS

1. What is the probability that a man has given birth?
2. A company has 5,000 employees. Approximately how many employees will have a birthday on any given day?
3. A state lottery spends 25% of its income on prizes, 20% on advertising, and 10% on salaries, and the rest goes to the state. How much should a person who spends $1,000 on scratch tickets each year expect to win?
4. Suppose the 10,000,000 people in a small country are all given a 99.9% accurate test for a disease that 10,000 of them have. What is the probability that a person who tests positive actually has the disease?
5. Raffle tickets for a car are being sold for $100 each. Only 350 tickets are sold. What is the probability of winning for a person who buys two tickets?
6. A person is going to draw a card randomly out of a deck of cards 500 times, shuffling the cards back together each time. How many times should she expect to draw out an ace or a face card?
7. There are 14 black socks, eight blue socks, and ten gray socks in a drawer. If one sock is drawn out randomly, what is the likelihood that the sock is black?
8. A person playing poker has four hearts and a club in his hand. If he discards the club and draws one new card, what is the probability of completing the flush (getting another heart)?
9. A person playing poker holds 3♥, 4♥, 5♠, 7♣, and Q♠. What is the probability of "drawing on an inside straight," discarding the queen, and getting a 6 to complete a straight?
10. What is the likelihood of rolling a 1 or a 2 on a die?
11. What is the probability of rolling doubles on two dice (both are the same number)?
12. What is the likelihood of rolling something other than a 1 five times in a row?
13. What is the probability of drawing a card from a deck and getting both a 10 and a black card?
14. What is the probability that a card drawn from a deck is both red and a diamond?
15. Suppose you are at a certain coffee shop about 5% of the time. If an employee works 30% of the time, what is the likelihood that you will both be there at the same time?
16. What is the probability of drawing two cards out of a deck and having both be aces?
17. A person playing poker is dealt three aces. If she discards the remaining two cards, what is the probability that one of the two cards she draws will be the fourth ace?
18. A particularly unsafe pickup truck has a 2% chance of exploding every day it is used. What is the probability that it can be used for 30 days without exploding?
19. What is the probability of rolling a die four times and getting at least one 5 or 6?
20. Over the winter, only three fish survive being frozen in a small pond. If the likelihood of a fish being male is 50%, what is the probability that the fish will be able to repopulate the pond (that there is at least one male and one female)?
21. How many different ways can the cards in a deck be ordered?
22. How many different ways can six paintings be arranged from left to right on a wall?
23. If ten people enter a contest, how many different ways can first, second, and third place be awarded?
24. Beside a door are 16 switches. A sign explains that the door will only open if the correct eight switches are flipped upward. How many different ways can eight of the 16 switches be chosen?
25. An artist wants to construct a perfume using three of 12 different essential oils. How many different ways can this be done?
26. There are ten girls and ten boys in a classroom. If three different students are chosen randomly, what is the probability that all three will be girls?
27. A coin is flipped ten times in a row. What is the probability that heads comes up four, five, or six times?

28. A box contains 20 lightbulbs, three of which are defective. If five bulbs are randomly taken out and tested, what is the probability that a defective bulb will be found?

29. To win a lottery, you must pick the correct six numbers, chosen from 1 to 49. What is the probability of winning?

30. A person has three diamonds in a five-card poker hand. If she discards the other two, what is the probability that the two new cards she draws will both be diamonds, resulting in a flush?

Answers

1. 0

2. 14

3. Approximately $250

4. 50%

5. $\dfrac{2}{350} \approx 0.6\%$

6. $500 \times \dfrac{4}{13} \approx 154$

7. $\dfrac{14}{32} = 43.75\%$

8. $\dfrac{9}{47} \approx 19\%$

9. $\dfrac{4}{47} \approx 8.5\%$

10. $\dfrac{2}{6} = 33\frac{1}{3}\%$

11. $\dfrac{6}{36} = 16\frac{2}{3}\%$

12. $\left(\dfrac{5}{6}\right)^5 \approx 40\%$

13. $\dfrac{1}{26} \approx 3.8\%$

14. 25%

15. 1.5%

16. $\dfrac{4}{52} \cdot \dfrac{3}{51} \approx 0.45\%$

17. $1 - \dfrac{46}{47} \cdot \dfrac{45}{46} = \dfrac{2}{47} \approx 4.3\%$

18. $(0.98)^{30} \approx 54.5\%$

19. $1 - \left(\dfrac{4}{6}\right)^4 = \dfrac{1,040}{1,296} \approx 80.2\%$

20. $1 - 2 \cdot \left(\dfrac{1}{2}\right)^3 = \dfrac{3}{4} = 75\%$

21. $52! \approx 8 \times 10^{67}$

22. $6! = 720$

23. $_{10}P_3 = 10 \times 9 \times 8 = 720$

24. $_{16}C_8 = 12,870$

25. $_{12}C_3 = 220$

26. $\dfrac{_{10}C_3}{_{20}C_3} = \dfrac{120}{1,140} \approx 10.5\%$

27. $\dfrac{_{10}C_4 + {_{10}C_5} + {_{10}C_6}}{2^{10}} = \dfrac{672}{1,024} = 65.625\%$

28. $1 - \dfrac{_{17}C_5}{_{20}C_5} = \dfrac{9,316}{15,504} \approx 60\%$

29. $\dfrac{1}{_{49}C_6} = \dfrac{1}{13,983,816}$

30. $\dfrac{_{10}C_2}{_{47}C_2} = \dfrac{45}{1,081} \approx 4.2\%$

CHAPTER 10

Statistics

The field of *statistics* encompasses the collection, depiction, and analysis of data.

The Depiction of Data

There are many ways to depict a large quantity of data. One way is to list out all the information. For example, a theatre that recorded the age of all attendees might come up with the following: 27, 27, 22, 47, 54, 28, 32, 38, 44, 50, 14, 68, 63, 66, 34, 36, 38, 27, 58, 35, 30, 24, 19, 33, 35, 44, and 45.

One way to simplify this data would be to group the data into ranges and record the number in each group with a *frequency chart*. For example, the information above could be put into age ranges that spanned ten years each, as shown in Table 10-1.

TABLE 10-1

AGE RANGE	0–10	11–20	21–30	31–40	41–50	51–60	61–70
Ages in range		14, 19	22, 24, 27, 27, 27, 28, 30	32, 33, 34, 35, 35, 36, 38, 38	44, 44, 45, 47, 50	54, 58	63, 66, 68
Frequency	0	2	7	8	5	2	3

Another way to depict this data would be with a *histogram*: a sequence of bars proportional in height to the frequency of each range, as shown in Fig. 10-1.

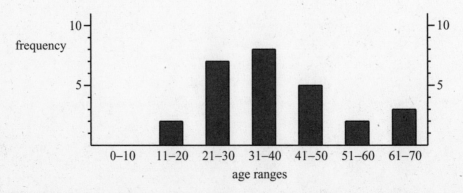

Fig. 10-1

Another way would be to calculate the percentage in each range. For example, there were 27 total people who attended the theatre, so the eight in the 31–40 range make up $\frac{8}{27} \approx 30\%$ of the whole. The percentage chart for the ongoing example is in Table 10-2.

TABLE 10-2

AGE RANGE	0–10	11–20	21–30	31–40	41–50	51–60	61–70
Percentage	0%	7%	26%	30%	19%	7%	11%

A *pie chart* depicts these percentages with proportional sectors (wedges) of a circle. For example, the 26% of the people in the 21–30 age range will be represented by a sector with an angle that measures $360° \times 0.26 = 93.6°$. The pie chart for the ongoing example is in Fig. 10-2.

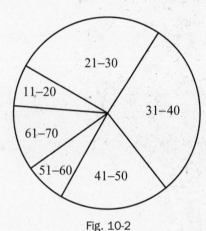

Fig. 10-2

SOLVED PROBLEMS

The Depiction of Data

1. A die is rolled 20 times, resulting in the following: 1, 4, 4, 1, 1, 4, 4, 2, 4, 6, 6, 5, 5, 1, 2, 3, 3, 3, 1, and 5.

    (a) Put this data into a frequency chart.
    (b) Depict this data with a histogram.
    (c) Make a chart depicting the percentage with which each number appeared.
    (d) Put this data into a pie chart.

2. A class of 28 students received the following grades on an exam: 60, 72, 100, 43, 79, 38, 82, 70, 52, 56, 87, 78, 64, 57, 96, 95, 47, 70, 68, 81, 61, 81, 51, 91, 38, 59, 89, and 77.

    (a) Put this data into a frequency chart, using the usual letter grades: 90–100=A, 80–89=B, 70–79=C, 60–69=D, and 0–59=F.
    (b) Depict this data with a histogram.
    (c) Make a chart depicting the percentage of the class which earned each letter grade.
    (d) Put this data into a pie chart.

3. The Fnord Motor Company claims that they have higher customer satisfaction than their competitors: X, Y, and Z. They illustrate their data with the chart in Fig. 10-3. What is wrong with this chart?

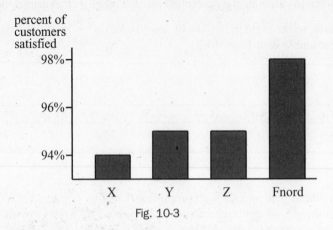

Fig. 10-3

Answers

1. (a) Shown in Table 10-3.

TABLE 10-3

DIE ROLL	1	2	3	4	5	6
Frequency	5	2	3	5	3	2

   (b) Shown in Fig. 10-4.

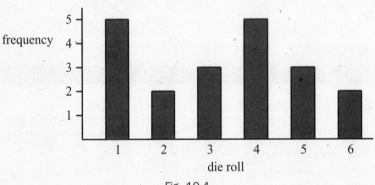

Fig. 10-4

   (c) Shown in Table 10-4.

TABLE 10-4

DIE ROLL	1	2	3	4	5	6
Frequency	25%	10%	15%	25%	15%	10%

(d) Shown in Fig. 10-5. The sectors representing 25% have angle $0.25 \times 360 = 90°$, the sectors representing 15% have angle $0.15 \times 360 = 54°$, and the sectors representing 10% have angle 36°.

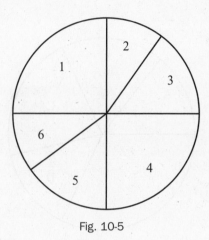

Fig. 10-5

2. (a) Shown in Table 10-5.

TABLE 10-5

GRADE	A	B	C	D	F
Frequency	4	5	6	4	9

(b) Shown in Fig. 10-6.

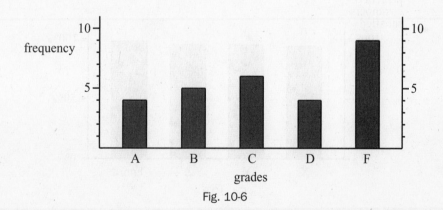

Fig. 10-6

(c) Shown in Table 10-6. Notice that because the numbers were rounded to the nearest whole percentage, their sum is not 100%.

TABLE 10-6

GRADE	A	B	C	D	F
Frequency	14%	18%	21%	14%	32%

(d) Shown in Fig. 10-7. The sector for A's measures $0.14 \times 360 = 50.4°$, or, to be more accurate,

$$\frac{4}{28} \times 360 \approx 51.43°.$$

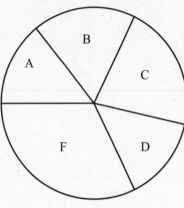

Fig. 10-7

3. If the customer satisfaction rates for X, Y, Z, and Fnord are really 94%, 95%, 95%, and 98% respectively, then the information in this chart is correct. However, this chart is made to look like a histogram, but it is not. In a histogram, the heights of the bars must be proportional to the values of the numbers. In this chart, the bar for Fnord looks more than twice as big as the bars for the other companies, even though 98% is only a small bit larger than 94% or 95%. This is a classic case of misusing statistics to deceive people. A true histogram for this data, illustrated in Fig. 10-8, shows that the difference between the company satisfaction rates is negligible.

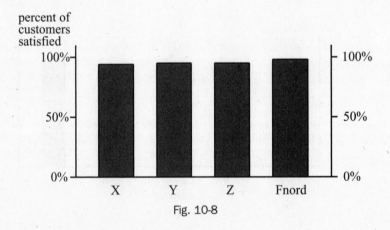

Fig. 10-8

Averages: Mean, Median, Midrange, and Mode

Given a list of numbers, there are several ways to estimate the average value.

The *mean* is obtained by adding up all the numbers and dividing by the number of data entries. The word *average* refers to the mean, unless specified otherwise.

The *median* is obtained by listing the numbers from smallest to largest and taking the one in the middle. If there are an even number of entries, there will be two in the middle. In this case, the median is the average of these two numbers.

The *midrange* is the average of the highest and lowest numbers.

The *mode* is the data value that occurs with the greatest frequency.

For example, suppose that a company has ten people who make $30,000 a year, five people who make $35,000, four who make $40,000, three who make $50,000, and one who makes $75,000.

These salaries range from 30,000 to $75,000, so the midrange salary is $\dfrac{30,000 + 75,000}{2} = \$52,500$.

This is not a very good measure of the average because it is higher than all but one of the salaries.

The most common salary is $30,000, so that is the mode. This is not a very good measure of the average because it is the lowest of all the salaries.

These 23 salaries, put in order and written in thousands, are as follows: 30, 30, 30, 30, 30, 30, 30, 30, 30, 30, 35, 35, 35, 35, 35, 40, 40, 40, 40, 50, 50, 50, and 75. The median salary will be the one in the middle, the 12th one: $35,000.

When the 23 salaries are added up, the sum is $860,000, so the mean is $\dfrac{860,000}{23} \approx \$37,391$.

In this case, the mean and median are quite close. The mean, however, can be strongly affected by an *extreme outlier*: a piece of data much higher or much lower than the rest. For example, suppose the person with the $75,000 salary also made $300,000 in stock options. If this total salary of $375,000 is included, then the mean will increase to $\dfrac{1,160,000}{23} \approx \$50,435$. It would be misleading to say that the average salary at the company was over $50,000 when only one person made that much. Furthermore, the midrange leaps to $\dfrac{30,000 + 375,000}{2} = \$202,500$, while the median and mode stay the same.

In general, the median is the most reliable measure of average because half of the data will be no more than this amount and the other half will be no less than it.

SOLVED PROBLEMS

Averages

1. Find the mean, median, midrange, and mode of these ages: 27, 22, 47, 28, 44, 14, 68, 33, and 45.
2. Find the mean, median, midrange, and mode of these exam grades: 60, 72, 100, 43, 79, 38, 82, 70, 52, 56, 87, 78, 64, 57, 96, 95, 47, 70, 68, 81, 61, 81, 51, 91, 38, 59, 89, and 77.
3. A professor teaches two sections of calculus: one with 25 students, and the second with 17 students. On an exam, the class average (mean) for the first class is 76%, and for the second class it is 68%. What is the mean grade for the two classes combined?
4. Suppose the company in this section's example has a labor dispute. Which sort of average would the striking workers quote? Which one would the management quote?

Answers

1. These numbers, put in order, are 14, 22, 27, 28, 33, 44, 45, 47, and 68. The middle is thus the fifth: 33, the median. Half of the numbers are ≤ 33, and half are ≥ 33. The sum of all is 328, so the mean is $36\frac{4}{9}$. This sort of average can turn out not to be any of the list of numbers. Each of the numbers on this list occurs only once, so there is no mode. Or rather, perhaps, they are all equally the mode. The midrange is $\dfrac{14 + 68}{2} = 41$.

2. The sum of these 28 numbers is 1,942, so the mean is $\dfrac{1,942}{28} \approx 69$. When arranged from smallest to largest, the result is 38, 38, 43, 47, 51, 52, 56, 57, 59, 60, 61, 64, 68, 70, 70, 72, 77, 78, 79, 81, 81, 82, 87, 89, 91, 95, 96, and 100. The median of these 28 numbers will be the mean of the 14th (70) and 15th (70); thus, the median is 70. These grades range from 38 to 100, so the midrange value is $\dfrac{38 + 100}{2} = 69$. All of the grades occur only once, except for three which occur twice. There are thus three different modes: 38, 70, and 81.

3. Even without the individual grades, we know that the sum of all the grades in the first class was $25 \times 76 = 1,900$, and for the second class, $17 \times 68 = 1,156$. Thus, the mean for the two classes combined will be $\dfrac{1,900 + 1,156}{25 + 17} \approx 72.8\%$

4. The striking workers would quote the lowest figure, the mode of $30,000, to generate the greatest sympathy for their cause. They would argue that this low salary was the most common among all those paid by the company. The management would quote the highest salary, the midrange of $52,500, to make the company look generous.

Standard Deviation

The *deviation* of a number in a set of data is the distance it is from the mean. The *standard deviation* is an estimate of the average deviation for a collection of data. For example, most adults are between five and six feet tall, so if the mean height is 5 feet, 6 inches, then most adults deviate by less than six inches. The standard deviation for salaries will be much more; a range of plus or minus $10,000 might be necessary to encompass a majority of the people.

To compute the standard deviation, add up the squares of all the deviations, divide by the number of pieces of data, and then take the square root. If there are n different pieces of data $(a_1, a_2, a_3, \ldots, a_n)$, with a mean of $\bar{x}$, then the standard deviation is $\sigma = \sqrt{\dfrac{(a_1 - \bar{x})^2 + (a_2 - \bar{x})^2 + \ldots (a_n - \bar{x})^2}{n}}$. In general, the majority of the numbers in the set will be between $\bar{x} - \sigma$ and $\bar{x} + \sigma$.

For example, suppose the weights of six friends are 135, 173, 180, 140, 160, and 154 pounds. The mean is $\dfrac{135 + 173 + 180 + 140 + 160 + 154}{6} = 157$ pounds. The number 135 has a deviation of $|135 - 157| = 22$. The number 173 has a deviation of $|173 - 157| = 16$. The standard deviation is $\sigma =$

$\sqrt{\dfrac{(135 - 157)^2 + (173 - 157)^2 + (180 - 157)^2 + (140 - 157)^2 + (160 - 157)^2 + (154 - 157)^2}{6}} \approx 16$. This indicates that the majority of the six weights are between $157 - 16 = 141$ and $157 + 16 = 173$ pounds.

The above formula is for the standard deviation of a complete set of data. If the information is only a sample of a much larger pool of data, then divide by $n - 1$ instead of n inside the square root. This $n - 1$ represents the *degrees of freedom*, the number of pieces of data we have to compare with the first piece of data. For example, if we wanted to figure out the average weight of American teenage males and randomly weighed 200 of them, we would divide by 199 instead of 200 when calculating the standard deviation of our sample.

SOLVED PROBLEMS

Standard Deviation

1. Over the course of a day, a museum is visited by eight school classes. The sizes of the classes are 18, 22, 12, 9, 27, 11, 29, and 16. What is the mean size of the classes? What is the standard deviation?

2. A company is interested in knowing how far its employees drive to work, so it chooses ten of them at random to ask. The responses are 30, 22, 18, 25, 38, 26, 11, 29, 4, and 24 miles. What are the mean and standard deviation of this sample?

3. An electrician has analyzed a year's worth of jobs and concluded that the mean job takes 90 minutes, with a standard deviation of 40 minutes. What does this mean?

Answers

1. The mean is $\bar{x} = \dfrac{18 + 22 + 12 + 9 + 27 + 11 + 29 + 16}{8} = 18$. The deviations are $|18 - 18| = 0, |22 - 18| = 4,$

    $|12 - 18| = 6, |9 - 18| = 9, |27 - 18| = 9, |11 - 18| = 7, |29 - 18| = 11,$ and $|16 - 18| = 2$. Thus, the standard deviation

    is $\sigma = \sqrt{\dfrac{0^2 + 4^2 + 6^2 + 9^2 + 9^2 + 7^2 + 11^2 + 2^2}{8}} \approx 7$.

2. The mean is $\bar{x} = \dfrac{30 + 22 + 18 + 25 + 38 + 26 + 11 + 29 + 4 + 24}{10} = 22.7$ miles. The deviations are

    $|30 - 22.7| = 7.3, |22 - 22.7| = 0.7, |18 - 22.7| = 4.7, |25 - 22.7| = 2.3, |38 - 22.7| = 15.3, |26 - 22.7| = 3.3,$
    $|11 - 22.7| = 11.7, |29 - 22.7| = 6.3, |4 - 22.7| = 18.7,$ and $|24 - 22.7| = 1.3$. Because this is a sample of the
    company's employees and not the list for the whole company, we divide by 9 instead of 10 in the standard
    deviation. Thus, the standard deviation is

    $$\sigma = \sqrt{\dfrac{7.3^2 + 0.7^2 + 4.7^2 + 2.3^2 + 15.3^2 + 3.3^2 + 11.7^2 + 6.3^2 + 18.7^2 + 1.3^2}{9}} \approx 9.7 \text{ miles.}$$

3. This means that the majority of her jobs take between $90 - 40 = 50$ minutes and $90 + 40 = 130$ minutes.

The Normal Curve and the Empirical Rule

When a histogram is drawn for a very large quantity of data, it often forms the shape of a *normal curve*, as illustrated in Fig. 10-9. Data for which this happens is called *normally distributed*. For examples, the grades on standardized tests, the sizes of manufactured products, and the measurements of men or women (height, weight, body temperature, etc.) are all normally distributed.

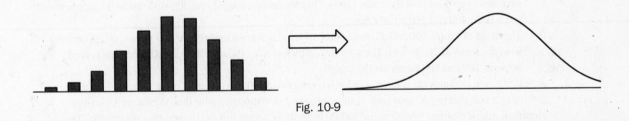

Fig. 10-9

There are other sorts of distributions. For example, if a die were rolled 1,000 times, all six possible outcomes will generally occur an equal number of times. This is called a *uniform distribution*, as shown in Fig. 10-10.

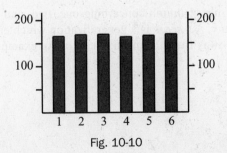

Fig. 10-10

The histogram of a normally distributed large collection of data can be approximated by a normal curve. The middle x-value of the curve is the mean, $\bar{x}$, of the data. As a rough rule of thumb (the *empirical rule*), 68% of the data will be within one standard deviation, σ, of the mean, as illustrated in Fig. 10-11(a). Similarly, 95% of the data will be within two standard deviations of the mean and 99.7% of the data will be within three standard deviations, as shown in Fig. 10-11(b) and 10-11(c).

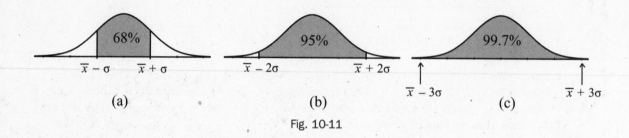

(a)　　　　　　　　　　(b)　　　　　　　　　　(c)

Fig. 10-11

SOLVED PROBLEMS

The Normal Curve and the Empirical Rule

1. Suppose that a standardized test is given to 100,000 high school students and the resulting scores are normally distributed. The mean score is 500 points and the standard deviation is 100 points. Approximately how many students scored (a) between 400 and 600, or (b) between 300 and 700?

2. A manufacturer purchases metal bolts from another company. The mean width of the bolts is 8 mm, and the standard deviation is 0.05 mm. If a bolt is wider than 8.15 mm or narrower than 7.85 mm, it cannot be used. In a shipment of 30,000 bolts, how many should the manufacturer expect to be too wide or too narrow to use?

Answers

1. (a) By the empirical rule, approximately 68% of the 100,000 students will score within one standard deviation (100 points) of the mean (500). This means that roughly 68,000 students will score between $500 - 100 = 400$ and $500 + 100 = 600$.

   (b) The range of scores 300 to 700 makes up two standard deviations (twice 100 points, or 200 points) from the mean of 500 points. Thus, by the empirical rule, about 95% or 95,000 students scored between 300 and 700 points on the exam.

2. A bolt that is 8.15 mm wide is three standard deviations ($\sigma = 0.05$) larger than the mean of 8 mm. Similarly, a bolt that is 7.85 mm wide is three standard deviations smaller than the mean. Using the empirical rule, we expect 99.7% of the 30,000 bolts to be within this range, and thus acceptable. This means that approximately $0.997 \times 30,000 = 29,910$ of the bolts should work. Only 90 of the bolts in the shipment should thus be too narrow or too wide to use.

Z Scores

The empirical rule was only a rule of thumb concerning one, two, and three multiples of the standard deviation σ away from the mean $\bar{x}$. A *z score* is a generalization which enables us to talk about any multiple z of the standard deviation away from the mean: $\bar{x} + z \cdot \sigma$. An example of a *z score chart* is shown in Table 10.7.

TABLE 10-7

Z score	−3.2	−3.0	−2.8	−2.6	−2.4	−2.2	−2.0	−1.8	−1.6	−1.4	−1.2
Area	.0007	.0013	.0026	.0047	.0082	.0139	.0228	.0359	.0548	.0808	.1151
Z score	−1.0	−0.8	−0.6	−0.4	−0.2	0	0.2	0.4	0.6	0.8	1.0
Area	.1587	.2119	.2743	.3446	.4207	.5000	.5793	.6554	.7257	.7881	.8413
Z score	1.2	1.4	1.6	1.8	2.0	2.2	2.4	2.6	2.8	3.0	3.2
Area	.8849	.9192	.9452	.9641	.9772	.9861	.9918	.9953	.9974	.9987	.9993

Each z score is associated with the percentage of the area up to $\bar{x} + z \cdot \sigma$ on the normal curve. For example, under the z score, 1.2 on the chart is area .8849. This means that 88.49% of the area under the normal curve is less than $\bar{x} + (1.2) \cdot \sigma$, as illustrated in Fig. 10-12.

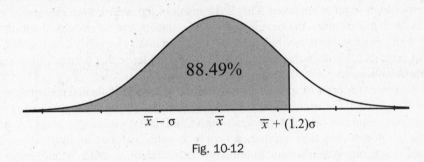

Fig. 10-12

Suppose, for example, that the mean IQ is $\bar{x} = 100$ and the standard deviation is $\sigma = 15$. This means that 88.49% of the population has an IQ less than or equal to $\bar{x} + (1.2) \cdot \sigma = 100 + (1.2)\ 15 = 118$. Only $100 - 88.49 = 11.51\%$ of the population thus has an IQ ≥ 118.

SOLVED PROBLEMS

Z Scores

1. Suppose a certain species of snake has a mean length of 24 inches with a standard deviation of five inches. How unusual would it be to find a snake of this species that was 35 inches long?
2. Suppose the average weight of an American adult is 160 pounds, with a standard deviation of 20 pounds. What percentage of American adults weigh less than 108 pounds?
3. Suppose the mean systolic blood pressure is 115 with a standard deviation of 15. What percentage of people have a systolic blood pressure between 103 and 136?

Answers

1. To calculate the z score associated with 35 inches, we solve $\bar{X} + z \cdot \sigma = 35$. Because $\bar{X} = 24$ and $\sigma = 5$, this means $24 + z \cdot 5 = 35$ so $z = \dfrac{11}{5} = 2.2$. Using the chart in Table 10-7, we see that this snake is bigger than or equal to 98.61% of all the snakes of this species. In general, a z score of two or greater (or less than −2) is considered unusual.
2. To find the z score, we solve $160 + z \cdot 20 = 108$, thus $z = −2.6$. According to Table 10-7, only 0.47% of American adults weigh less than or equal to 108 pounds.
3. The z score related to 136 is $115 + z \cdot 15 = 136$, thus $z = 1.4$. According to Table 10-7, 91.92% of the population has a systolic blood pressure ≤ 136. The z score related to 103 is $115 + z \cdot 15 = 103$, thus $z = −0.8$. According to Table 10-7, 21.19% of the population has systolic blood pressure ≤ 103. We thus conclude that $91.92 − 21.19 = 70.73\%$ of the population has systolic blood pressure between 103 and 136.

Collecting Statistics

When data is obtained for an entire *population* (group under study), it is only a matter of arithmetic to calculate the mean and standard deviation. Usually this is impossible: for example, no one knows the opinions of all American voters or the sizes of all bears in the wild. The only option is to *sample* (select) some of the population to study.

It is very important that the sample be *random*: that each object in the population has an equal chance of being selected. A selection of people leaving a mosque, for example, will likely have different views on alcohol than a selection leaving a night club. Neither selection will probably reflect the country as a whole. A common method for random sampling is to call numbers taken blindly from phone books or national databases of phone numbers. Even this method has a problem; the population for this method consists only of people whose phones are listed in the phone book (which does not include cell phones) and who answer phone surveys.

Even if data from a population is not normally distributed, random samples taken from the data will form a normal curve if enough samples are taken. This is the result of the *central limit theorem*, a great achievement in statistics. With this theorem, the *margin of error* of a statistic can be calculated within an interval of confidence. If a well-conducted poll concludes that 64% of the country supports a certain bill, for example, with a margin of error of ±3%, then there is a high level of confidence that between 61% and 67% of the country supports the bill.

There are many places for error to appear when collecting statistics. If slanted questions are asked, then the responses will be biased. For example, "Do you agree with Candidate A's plan to ruin the economy?" will generate different responses than "Do you agree with Candidate A's economic plan?" Another potential source of trouble is self-reporting bias: relying on people to judge and measure themselves. Many people have an interest in exaggerating their wealth, minimizing their infidelities, and generally making themselves look better, for example. Tax records, for example, give a more accurate reflection of a person's wealth than what he or she will tell you.

When conducting a scientific study, it is important to do everything possible to avoid bias. To examine the usefulness of a medical treatment, for example, some of the patients must be given the treatment (the *experimental group*) and others must not be given the treatment (the *control group*). Some patients get better just from knowing that they are receiving a treatment, however, and so it is important to give a *placebo* (fake drug or treatment) to the control group. A study is *blind* when the patients do not know whether they are receiving the treatment or not. Even better is a *double-blind* study, where the doctors and researchers do not know which treatments contain the real drugs and which ones are placebos. The doctors might treat the two groups of patients differently if the study was not double-blind, and this could affect the results.

SOLVED PROBLEMS

Collecting Statistics

1. Suppose a poll is taken on "death penalty: yes or no?" The poll results in 52% saying "yes," with a 3% margin of error. Does this mean that a majority agree with the death penalty?
2. Suppose a news article claims that 60% of all Americans believe that there are aliens on earth. What questions ought to be asked before believing this statistic?
3. What is the main challenge in clinically testing the effectiveness of acupuncture (therapy involving needles)?

Answers

1. This means that anywhere between 49% and 55% agree with the death penalty. Because the number could very possibly be below 50%, it cannot be said that a majority agree with the death penalty.
2. Who conducted this study? How many people were asked? How were these people selected? What was the exact wording of the question? What is the margin of error?

3. While it is easy to make a fake pill to serve as a placebo in a drug study, it is difficult to design a placebo for acupuncture. The patients in the study would have to be treated in such a way that they could not tell if they had been stuck with needles as part of an acupuncture treatment or not.

SUPPLEMENTAL PROBLEMS

1. The chefs in a town are asked to pick their favorite ingredient from garlic (G), cilantro (C), salt (S), pepper (P), or bacon (B). The results are C, P, B, S, C, S, S, C, C, C, B, G, B, C, P, S, S, B, S, and S. Convey this data in (a) a frequency chart, (b) a histogram, (c) a chart of percentages, and (d) a pie chart.
2. The weights of the dogs at a kennel are 72, 29, 89, 74, 24, 62, 7, 36, 48, 41, 22, 12, 37, 64, 56, 27, 31, 44, 36, 29, 18, 45, 63, 33, and 28 pounds. Find the (a) mean, (b) median, (c) midrange, and (d) mode of this data. Calculate the (e) standard deviation. Group this data into ranges of 10 pounds each (0–10, 11–20, etc.), and use this to draw (f) a frequency chart, (g) a histogram, (h) a percentage chart, and (i) a pie chart.
3. The runners on a track team have recorded their best times for running a mile: 250, 232, 218, 245, 260, 220, 223, 225, 214, and 235 minutes. Find the (a) mean, (b) median, (c) mode, (d) midrange, and (e) standard deviation of these times.
4. The mean age of the people in room 1 is 47 years. The mean age of the people in room 2 is 32 years. If there are 20 people in room 1 and 35 people in room 2, then what is the mean age of all the people in rooms 1 and 2 put together?
5. A study finds that the average family has 2.1 children. Is this a mean, median, midrange, or mode?
6. What are the mean and standard deviation of these numbers: 16, 18, 25, 30, 19, 22, 14, 21, 28, 22, 16, and 20?
7. Suppose the mean length of a shower is eight minutes, with a standard deviation of two minutes. What does the empirical rule say about the number of showers (a) between six and ten minutes long, (b) between four and 12 minutes long, and (c) between two and 14 minutes long? You may suppose that the lengths of showers are normally distributed.
8. Suppose the mean weight of a cat is 12 pounds, with a standard deviation of 3 pounds. Suppose further that you have randomly collected 5,000 cats and put them in a pile. Assuming that the weights of cats are normally distributed, what does the empirical rule say about the number of cats in your pile that weigh between 9 and 15 pounds?
9. Suppose the mean height of adult males is 5 feet, 8 inches tall and that the standard deviation is 3 inches. Use the z score chart in Table 10-7 to calculate the percentage of men shorter than 5 feet, 2 inches tall.
10. Suppose the mean weight of a pumpkin is 27 pounds, with a standard deviation of five pounds. In a pumpkin patch of 1,000 pumpkins, how many would you expect to be at least 40 pounds? (Use the chart in Table 10-7.)
11. Suppose the mean shoe size for women is 8, with a standard deviation of one. If a slipper is designed to fit any size from 5 to 10, what percentage of women will this slipper fit?
12. Suppose the mean IQ is 100 and the standard deviation is 15. If a person is randomly selected, what is the probability that he or she has an IQ of at least 127?
13. A study has people eat a bowl of cereal for breakfast each morning. At the end of the study, 50% of the people participating have lost weight. The cereal box then proudly proclaims this result. What is the smallest number of people who could have participated in this study?

Answers

1. (a) Shown in Table 10-8.

TABLE 10-8

INGREDIENT	GARLIC	CILANTRO	SALT	PEPPER	BACON
Frequency	1	6	7	2	4

(b)　Shown in Fig. 10-13.

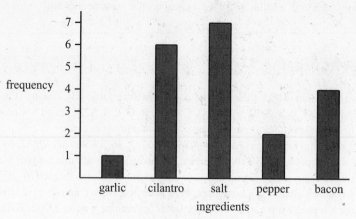

Fig. 10-13

(c)　Shown in Table 10-9.

TABLE 10-9

INGREDIENT	GARLIC	CILANTRO	SALT	PEPPER	BACON
Frequency	5%	30%	35%	10%	20%

(d)　Shown in Fig. 10-14.

Fig. 10-14

2.　(a) 41.08, (b) 36, (c) 48, (d) 29 and 36, (e) ≈ 20.2, and (f) shown in Table 10-10.

TABLE 10-10

WEIGHT	0–10	11–20	21–30	31–40	41–50	51–60	61–70	71–80	81–90
Frequency	1	2	6	5	4	1	3	2	1

(g)   Shown in Fig. 10-15.

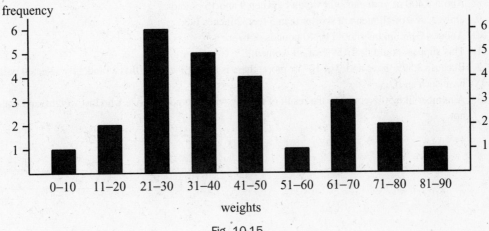

Fig. 10-15

(h)   Shown in Table 10-11.

TABLE 10-11

WEIGHT	0–10	11–20	21–30	31–40	41–50	51–60	61–70	71–80	81–90
Frequency	4%	8%	24%	20%	16%	4%	12%	8%	4%

(i)   Shown in Fig. 10-16.

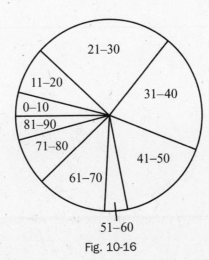

Fig. 10-16

3.   (a) 232.2, (b) 228.5, (c) no mode, (d) 237, and (e) ≈14.4

4. $\dfrac{2{,}060}{55} = 37\frac{5}{11}$

5.   This is a mean, because every family has a natural number of children, and the median, midrange, and mode of natural numbers will either be natural or have fractional part 0.5.

6.   Mean ≈ 20.9, standard deviation ≈ 4.7

7. (a) About 68% of all showers are of this length, (b) about 95% of all showers are of this length, and (c) about 99.7% of all showers are of this length.
8. About 3,400 of your cats will weigh between 9 and 15 pounds.
9. Only 2.28% of all men are shorter than 5 feet, 2 inches tall.
10. About 4.7 pumpkins should be 40 pounds or over.
11. This slipper should fit 97.59% of all women.
12. Because 3.59% have an IQ of 127 or more, there is a 3.59% chance that a randomly selected person will have such an IQ.
13. A number like 50% could be the result of a study with two people: one who lost weight, and one who did not.

CHAPTER 11

Weighted Voting

When a member of a committee wants the group to collectively do something, he or she must submit a *motion*, a clear description of the idea. After the group discusses and perhaps modifies the proposal, they vote on it. If the motion gets a predetermined number of votes, called the *quota*, it passes and the committee then acts upon it. Otherwise, the motion fails and the committee must find something else to do.

Sometimes the members of a committee have different numbers of votes. The *weight* of a member is the number of votes he or she has. For example, the weight of a participant at a corporate board meeting is the number of shares of stock that person owns. Such a situation is called a *weighted voting system*.

Describing Weighted Voting Systems

Mathematically, a weighted voting system can be described by listing the quota, followed by the weights of all the voters (usually called *players*). As a convention, the weights are listed in decreasing order. We will suppose that the first player is called A, the second is player B, etc. For example, [5: 3, 2, 1, 1, 1, 1] represents a weighted voting system for six players where it takes five votes to pass a motion. Player A has three votes, player B has two votes, and players C, D, E, and F have one vote each.

The quota cannot be more than the total of all the players' votes; for example, [50: 5, 2, 2] is not a valid weighted voting system, because no motion could ever pass. Even if A, B, and C all voted "yes" to a motion, they would only have $5+2+2=9$ votes, well short of the 50-vote quota.

The quota must also be over half the total of all the votes. Otherwise, half the votes could be "yes," the other half could be "no," and both would meet the quota. For example, [7: 5, 4, 4, 3, 3] is not a valid weighted voting system because the quota of seven is less than half the total of all the votes: $5+4+4+3+3=19$.

SOLVED PROBLEMS

Describing Weighted Voting Systems

Describe the number of players, their weights, and the quota of each valid weighted voting system. If the system is not valid, explain why.

1. [8: 2, 2, 2, 1, 1, 1, 1]
2. [15: 5, 3, 3, 2]
3. [6: 3, 2, 1]
4. [10: 12, 3, 3, 1]
5. [3: 2, 1, 1, 1, 1]
6. [10: 8, 4, 2, 2]

Answers

1. There are seven players; three have two votes each, and four have one vote each. It requires a quota of at least eight votes to pass a motion.

2. This is not a valid weighted voting system because the sum of all the players' votes is less than the quota: $5+3+3+2=13<15$.

3. There are three players. Player A has three votes, B has two votes, and C has one vote. Because the quota is six, it will take everyone's votes to pass a motion.

4. There are four players: player A has 12 votes, B has three votes, C has three votes, and D has one vote. The quota is ten.

5. This is not a valid weighted voting system because the quota of three is not more than half the total of all the players' votes: $2+1+1+1+1=6$. If players A and B vote "yes" while C, D, and E vote "no," then both groups would meet the quota.

6. This is a valid weighted voting system for four players. Player A has eight votes, B has four votes, and players C and D have two votes each. The quota is ten.

Dictators, Dummies, and Veto Power

A *dictator* is a player in a weighted voting system who can pass a motion without help from anyone else. This happens when the player's weight is at least as much as the quota. For example, in [10: 12, 5, 1, 1, 1] player A is a dictator because $12 \geq 10$.

A *dummy* is a player whose vote never matters. For example, in [4: 2, 2, 1] a motion will pass if A and B both vote for it. If either A or B votes against the motion, it will fail. It makes no difference how C votes. For this reason, C is a dummy.

A player can *veto* a motion if he or she can stop it by voting "no." This happens when the combined weight of all the other players does not reach the quota. For example, in [6: 5, 3, 1] player A has veto power because B and C have only four votes together, not enough for the quota. Player B does not have veto power because A and C can pass a motion with their combined six votes, even if B votes "no."

SOLVED PROBLEMS

Recognizing Dictators, Dummies, and Veto Power

Identify any dictators, dummies, and people with veto power in the following weighted voting systems:

1. [16: 10, 6, 4, 2, 2]
2. [8: 8, 2, 2, 1]
3. [6: 3, 3, 3, 1]
4. [11: 5, 5, 1, 1, 1]
5. [8: 4, 2, 1, 1]
6. [4: 1, 1, 1, 1]

Answers

1. None of the players have the 16 votes necessary to be a dictator. Player A has veto power, however, because the rest of the players have $6+4+2+2=14$ votes, not enough to meet the quota. There are no dummies, because even D or E can make a difference if A and C vote "yes" and B votes "no."

2. Player A is a dictator who can pass any motion without help. Player A also has veto power because the other players have only five combined votes. Because player A alone determines if a motion passes or fails, all the other players are dummies.

3. There is no dictator with six or more votes. Player A does not have veto power because B, C, and D together have seven votes, more than enough to pass a motion without A. Because A has as much weight as

anyone, no one has veto power. Motions require two of the first three players (A, B, and C) to vote for it, and fail otherwise. Thus, D is a dummy.

4. Player A has veto power because the others have only $5+1+1+1=8<11$. Player B has the same weight as A, and thus also has veto power. There are no dummies or dictators.

5. There are no dummies or dictators here. Instead, everyone has veto power. Without D's one vote, the others have only $4+2+1=7<8$. This is called a *unanimous voting system* because motions only pass if everyone votes "yes."

6. This is another unanimous voting system for four people: everyone has veto power, no one is a dictator, and there are no dummies.

Setting Up a Weighted Voting System

To devise a weighted voting system with a dictator, give the dictator more votes than all the other players combined and set the quota equal to the dictator's weight. For example, if there are seven players, give one vote each to six of them, give seven votes to the dictator, and set the quota at seven. Thus, [7: 7, 1, 1, 1, 1, 1, 1] is a weighted voting system for seven players that has a dictator.

To give some players veto power, give them more votes than all of the nonveto players combined. Then set the quota to at least the total weight of those with veto. For example, suppose we want a weighted voting system for ten players: three with veto power and seven without. Give each of the seven players without veto power one vote each. Give each of the three with veto power eight votes each. Any quota between 24 (the combined weight of those with veto) and 31 (the combined weight of all players) will work. In [24: 8, 8, 8, 1, 1, 1, 1, 1, 1, 1], there are three players with veto and seven dummies. In [27: 8, 8, 8, 1, 1, 1, 1, 1, 1, 1], there are three players with veto and it takes at least six players to pass a motion.

SOLVED PROBLEMS

Setting Up Weighted Voting Systems

Devise a weighted voting system with the following requirements:

1. Twelve players, including one dictator
2. Five players, one with veto power, and no dummies
3. Six players, two with veto power, and four dummies
4. Eight players, four with veto power, where it takes at least six people to pass a motion

Answers

There are many possible answers, but here are some:

1. [12 : 12, 1, 1, 1, 1, 1, 1, 1, 1, 1, 1, 1]
2. [6 : 5, 1, 1, 1, 1], though the quota could also be seven or eight.
3. [10 : 5, 5, 1, 1, 1, 1]
4. [22: 5, 5, 5, 5, 1, 1, 1, 1]

The Banzhaf Power Index

A dictator is said to have 100% of the power to decide the outcome of a vote. A dummy has 0% of the decision-making power. More generally, a method called the *Banzhaf Power Index* can be used to calculate the relative power of each player in a weighted voting system.

To compute the Banzhaf Power Index:

1. List all the combinations of players who together could pass a motion. These are the groups of players whose combined weight is at least the quota. Such groups are called *winning coalitions*.
2. Take a player from one of the winning coalitions and subtract his or her weight from the coalition total. If the remaining players do not meet the quota, the subtracted player is called *critical* to the coalition's success. Underline that player's name in that particular coalition.
3. Repeat step 2 for every player in every winning coalition.
4. A player's Banzhaf Power Index is the number of times he or she is critical (underlined), divided by the number of times anyone is critical.

As an example, let us calculate the Banzhaf Power Index for the three players in [4: 3, 2, 1]. The winning coalitions are {A, B} (with a combined weight of five), {A, C} (combined weight of four), and {A, B, C} (combined weight of six). No other group of players reaches the quota of four with their combined weight.

In winning coalition {A, B}, player A is critical because if A's weight is subtracted, B alone only has two votes, not enough for the quota. Similarly, B is critical because A alone cannot pass a motion. In coalition {A, C}, both players are again critical. In {A, B, C}, only A is critical. Players B and C cannot reach the quota by themselves. If B is subtracted, however, {A, C} remains a winning coalition.

The list of winning coalitions with critical players underlined is thus: {A, B}, {A, C}, and {A, B, C}. There are five instances where someone is critical. Player A is critical in three of them, so player A's Banzhaf Power Index is $\dfrac{3}{5} = 60\%$. Player B is critical in only one instance, so B's power index is $\dfrac{1}{5} = 20\%$.

Similarly, C is critical only once, and thus C's power is $\dfrac{1}{5} = 20\%$.

According to the Banzhaf Power Index, player A has 60% of the power and players B and C each have 20% of the power. This is perhaps surprising because B has twice as many votes as C. However, B's vote is only critical in one situation; the same is true of C.

Here are a few tips on finding winning coalitions and critical players. A player with veto power will be in every winning coalition and will always be critical. A player not in a winning coalition can be added to form a new winning coalition, but will never be critical in the new coalition. For example, in [7: 4, 3, 3, 2], the coalition {A, B} has seven combined votes and thus is a winning coalition. Any combination of players C and D can be added to form new winning coalitions: {A, B, C}, {A, B, D}, and {A, B, C, D}. In these last three coalitions, neither C nor D will ever be critical. In general, a player is critical to a winning coalition if the remaining players do not form a winning coalition.

SOLVED PROBLEMS

Calculating the Banzhaf Power Index

Calculate the Banzhaf Power Index for each player in the following weighted voting systems:

1. [3: 2, 1, 1, 1]
2. [16: 8, 4, 2, 1, 1]
3. [6: 4, 3, 2, 1]
4. [6: 3, 3, 1, 1, 1]
5. [10: 3, 3, 3, 1, 1]

Answers

1. The winning coalitions (with critical players underlined) are {A, B}, {A, C}, {A, D}, {A, B, C}, {A, B, D}, {A, C, D}, {A, B, C, D}, and {B, C, D}. There are 12 instances where a player is critical. Player A is critical in six situations, so the Banzhaf Power Index of A is $\dfrac{6}{12} = 50\%$. Players B, C, and D are each critical in two situations, so the Power Index of each is $\dfrac{2}{12} = 16\dfrac{2}{3}\%$.

2. The only winning coalition is {A, B, C, D, E}, and everyone is critical. Each player is critical once out of the five critical instances, and thus the Power Index of each is $\frac{1}{5} = 20\%$. In every unanimous voting system, the players have equal power.

3. The winning coalitions are {A, B}, {A, B, C}, {A, B, D}, {A, B, C, D}, {A, C}, {A, C, D}, and {B, C, D}. There are 12 situations where someone is critical. The power of player A is $\frac{5}{12} = 41\frac{2}{3}\%$. The power of player B is $\frac{3}{12} = 25\%$. The power of player C is also $\frac{3}{12} = 25\%$. The power of player D is $\frac{1}{12} = 8\frac{1}{3}\%$.

4. The winning coalitions are {A, B}, {A, B, C}, {A, B, D}, {A, B, E}, {A, B, C, D}, {A, B, C, E}, {A, B, D, E}, {A, B, C, D, E}, {A, C, D, E}, and {B, C, D, E}. The power of players A and B is $\frac{8}{22} = 36\frac{4}{11}\%$ each. The power of players C, D, and E is $\frac{2}{22} = 9\frac{1}{11}\%$ each.

5. The winning coalitions are {A, B, C, D}, {A, B, C, E}, and {A, B, C, D, E}. The power of players A, B, and C is $\frac{3}{11} = 27\frac{3}{11}\%$ each. The power of players D and E is $\frac{1}{11} = 9\frac{1}{11}\%$ each.

SUPPLEMENTAL PROBLEMS

1. For each weighted voting system, either describe the number of players, their weights, and the quota, or else explain why the system is not valid:

   (a) [6: 3, 3, 1]
   (b) [9: 10, 4, 1, 1]
   (c) [25: 9, 7, 3, 3, 1]
   (d) [35: 10, 10, 10, 5, 5, 5]
   (e) [6: 3, 3, 2, 2, 2]
   (f) [40: 20, 14, 14, 12, 9, 5]
   (g) [20: 5, 5, 3, 2, 1, 1, 1, 1, 1]
   (h) [40: 15, 12, 11, 11, 9, 9, 7, 7]

2. Identify any dictators, dummies, and players with veto power in the following systems:

   (a) [4: 4, 2, 1]
   (b) [6: 5, 3, 2, 1]
   (c) [14: 8, 4, 2, 1, 1]
   (d) [10: 9, 2, 2, 2, 2]
   (e) [11: 12, 3, 3, 1, 1, 1]
   (f) [17: 7, 5, 3, 2]
   (g) [10: 5, 5, 5, 2, 1, 1]
   (h) [18: 4, 4, 4, 4, 1, 1, 1, 1, 1]

3. Explain why holding 51% of a company's stock gives you 100% control of the company.
4. Why is it impossible for there to be two dictators in a weighted voting system?
5. Why does a dictator automatically have veto power?
6. Why does the presence of a dictator make all the other players dummies?
7. Devise a weighted voting system for the following situations:

   (a) Eight people, all with equal power, where it takes three-quarters of all people to pass a motion
   (b) A unanimous system for six players
   (c) Seven players, four with veto power and three dummies
   (d) Five dummies and one dictator
   (e) Ten players, two with veto, where it takes at least seven people to pass a motion
   (f) Twelve players, three with veto, where it takes at least five people to pass a motion

8. There are 15 countries on the United Nations Security Council. Five of them have veto power (China, France, Great Britain, Russia, and the United States). A resolution can only be passed with nine countries supporting it. Express this as a weighted voting system.

9. A group of seven friends wants a system for voting on things to do. They want at least five people to agree on an idea before it happens. The only one with a car insists on having veto power. How can a weighted voting system be devised for them?

10. For each of the following weighted voting systems, list all the winning coalitions and calculate the Banzhaf Power Index for each player:

   (a) [3: 2, 1, 1]
   (b) [5: 5, 1, 1, 1]
   (c) [8: 5, 1, 1, 1, 1]
   (d) [10: 5, 5, 3, 3]
   (e) [8: 5, 3, 2, 2]
   (f) [10: 4, 4, 4, 2, 2]
   (g) [8: 5, 3, 2, 1, 1]

11. In ancient Sparta, the king had two votes and all other men had one. Suppose a simple majority (any number of votes more than half) was enough to pass a motion. Write out the weighted voting system and calculate the power of the king when he made decisions with (a) one, (b) two, (c) three, and (d) four other people.

12. In a weighted voting system, the players with veto will all have the same Banzhaf Power Index. Explain.

Answers

1. (a) There are three players, two with three votes and one with one vote, and the quota is six.
    (b) The quota is nine, and there are four players: A has ten votes, B has four votes, C has one vote, and D has one vote.
    (c) Not valid—the quota is too high.
    (d) The quota is 35, and there are six players: three with ten votes, and three with five votes.
    (e) Not valid—the quota is too low.
    (f) The quota is 40, and there are six players: A has 20 votes, B has 14 votes, C has 14 votes, D has 12 votes, E has 9 votes, and F has 5 votes.
    (g) The quota is 20, and there are nine players: two have five votes, one has three votes, one has two votes, and five have one vote each.
    (h) Not valid—the quota is too low.

2. (a) Player A is a dictator with veto power, and players B and C are dummies.
    (b) There are no dictators, dummies, or people with veto power.
    (c) Players A and B have veto power.
    (d) Player A has veto power.
    (e) Player A is a dictator with veto power, and players B, C, D, E, and F are dummies.
    (f) Every player has veto power.
    (g) Players D, E, and F are all dummies.
    (h) Players A, B, C, and D all have veto power.

3. A person with 51% of a company's stock is a dictator when the company votes by shares and the quota is one vote more than half, as is usually the case.

4. Two dictators in a system would mean that two players had at least as many shares as the quota. This means that the quota is not more than half the weight of all the players, and thus the system is not valid.

5. A dictator has at least as many votes as the quota. The quota is at least half of the total weight of all the players. The total weight of all the players except the dictator must thus be less than the quota. This means that the dictator has veto power.

6. A dictator can pass or reject a motion without help from any of the other players. This means that the votes of the others cannot make a difference; thus, they are dummies.

7. (a) [6: 1, 1, 1, 1, 1, 1, 1, 1]
   (b) [6 : 1, 1, 1, 1, 1, 1]
   (c) [16 : 4, 4, 4, 4, 1, 1, 1]
   (d) [6: 6, 1, 1, 1, 1, 1]
   (e) [23 : 9, 9, 1, 1, 1, 1, 1, 1, 1, 1]
   (f) [32 : 10, 10, 10, 1, 1, 1, 1, 1, 1, 1, 1, 1]
8. [59 : 11, 11, 11, 11, 11, 1, 1, 1, 1, 1, 1, 1, 1, 1, 1]
9. [11 : 7, 1, 1, 1, 1, 1, 1]

10. (a) Winning coalitions: {$\underline{A}$, $\underline{B}$}, {$\underline{A}$, $\underline{C}$}, and {$\underline{A}$, B, C}. The power of A is $\frac{3}{5}$ = 60%, and the power of B and C is $\frac{1}{5}$ = 20% each.

   (b) Winning coalitions: {$\underline{A}$}, {$\underline{A}$, B}, {$\underline{A}$, C}, {$\underline{A}$, D}, {$\underline{A}$, B, C}, {$\underline{A}$, B, D}, {$\underline{A}$, C, D}, and {$\underline{A}$, B, C, D}. The power of A is 100%. Players B, C, and D have 0% of the power each.

   (c) Winning coalitions: {$\underline{A}$, $\underline{B}$, $\underline{C}$, $\underline{D}$}, {$\underline{A}$, $\underline{B}$, $\underline{C}$, $\underline{E}$}, {$\underline{A}$, $\underline{B}$, $\underline{D}$, $\underline{E}$}, {$\underline{A}$, $\underline{C}$, $\underline{D}$, $\underline{E}$}, and {A, B, C, D, E}. The power of A is $\frac{5}{17}$ = $29\frac{7}{17}$%. The power of B, C, D, and E is $\frac{3}{17}$ = $17\frac{11}{17}$% each.

   (d) Winning coalitions: {$\underline{A}$, $\underline{B}$}, {$\underline{A}$, $\underline{B}$, C}, {$\underline{A}$, $\underline{B}$, D}, {A, B, C, D}, {$\underline{A}$, $\underline{C}$, $\underline{D}$}, and {$\underline{B}$, $\underline{C}$, $\underline{D}$}. The power of A and B is $\frac{4}{12}$ = $33\frac{1}{3}$% each. The power of C and D is $\frac{2}{12}$ = $16\frac{2}{3}$% each.

   (e) Winning coalitions: {$\underline{A}$, $\underline{B}$}, {$\underline{A}$, $\underline{B}$, C}, {$\underline{A}$, $\underline{B}$, D}, {$\underline{A}$, B, C, D}, and {$\underline{A}$, $\underline{C}$, $\underline{D}$}. The power of A is $\frac{5}{10}$ = 50%. The power of B is $\frac{3}{10}$ = 30%. The power of C and D is $\frac{1}{10}$ = 10% each.

   (f) Winning coalitions: {$\underline{A}$, $\underline{B}$, $\underline{C}$}, {A, B, C, D}, {A, B, C, E}, {A, B, C, D, E}, {$\underline{A}$, $\underline{B}$, $\underline{D}$}, {$\underline{A}$, $\underline{B}$, $\underline{E}$}, {$\underline{A}$, $\underline{B}$, D, E}, {$\underline{A}$, $\underline{C}$, $\underline{D}$}, {$\underline{A}$, $\underline{C}$, $\underline{E}$}, {$\underline{A}$, $\underline{C}$, D, E}, {$\underline{B}$, $\underline{C}$, $\underline{D}$}, {$\underline{B}$, $\underline{C}$, $\underline{E}$}, and {$\underline{B}$, $\underline{C}$, D, E}. The power of A, B, and C is $\frac{7}{27}$ = $25\frac{25}{27}$% each. The power of D and E is $\frac{3}{27}$ = $11\frac{1}{9}$% each.

   (g) Winning coalitions: {$\underline{A}$, $\underline{B}$}, {$\underline{A}$, $\underline{B}$, C}, {$\underline{A}$, $\underline{B}$, D}, {$\underline{A}$, $\underline{B}$, E}, {$\underline{A}$, $\underline{B}$, C, D}, {$\underline{A}$, $\underline{B}$, C, E}, {$\underline{A}$, $\underline{B}$, D, E}, {$\underline{A}$, B, C, D, E}, {$\underline{A}$, $\underline{C}$, $\underline{D}$}, {$\underline{A}$, $\underline{C}$, $\underline{E}$}, and {$\underline{A}$, $\underline{C}$, D, E}. The power of A is $\frac{11}{21}$ = $52\frac{8}{21}$%. The power of B is $\frac{5}{21}$ = $23\frac{17}{21}$%. The power of C is $\frac{3}{21}$ = $14\frac{2}{7}$%. The power of D and E is $\frac{1}{21}$ = $4\frac{16}{21}$% each.

11. (a) [2 : 2, 1], so the king had 100% of the power.
    (b) [3 : 2, 1, 1], so the king had 60% of the power.
    (c) [3 : 2, 1, 1, 1], so the king had 50% of the power.
    (d) [4 : 2, 1, 1, 1, 1], so the king had $\frac{10}{26}$ = $38\frac{6}{13}$% of the power.

12. A player with veto power will be in every winning coalition because no vote could be carried without his or her vote. This means that the player is critical in every winning coalition. Thus, the power of any veto-wielding player is the number of winning coalitions divided by the number of times a player is critical: the same is true for each player with veto power.

CHAPTER 12

Voting Methods

In this chapter, we will examine various *voting methods*, ways to decide the winner of an election. To compare the different methods, we will discuss what it means for a voting method to decide an election unfairly via various *fairness criteria*.

Majorities and Pluralities

It is easiest to decide a *unanimous* election where all the voters choose the same candidate. This is very rare, but can occur when there are few voters. Sometimes tyrants win unanimous elections, but only by eliminating all other candidates and threatening the voters with violence.

The next easiest elections to decide occur when one candidate wins a *majority* (more than half) of the votes. Such a person is called the *majority candidate*. The *majority fairness criterion* insists it would be unfair for a majority candidate to lose the election. When there are only two candidates, one of them will win a majority unless there is a tie.

When there are three or more candidates, it is possible for no candidate to win a majority. For example, suppose the students of a high school senior class vote for class color, as given in Table 12-1.

TABLE 12-1

Votes	21	47	17	24	39	32
Color	Black	Red	Orange	Green	Blue	Purple

There were a total of 180 votes cast, so more than 90 would be necessary for a majority. Because no color won 91 or more votes, there is no majority candidate.

In the United States, the *plurality voting method* is most often used: the candidate with the most votes wins. In Table 12-1, the color red would win because it received a plurality of the votes.

SOLVED PROBLEMS

Majorities and Pluralities

1. Suppose 24,000 people vote in an election. (a) How many votes are needed for a unanimous victory? (b) How many votes are necessary for a majority victory? (c) How many votes are necessary for a plurality victory when four candidates compete?
2. Suppose 100,000 voters will use the plurality voting method to find a winner from eight candidates. (a) What are the fewest votes with which a candidate could win? (b) What percentage of the total vote would this be? (c) What is the largest number of votes with which a candidate could win?

3. If the plurality voting method is used, could a majority candidate lose an election?
4. Why is the voting in the United States sometimes called *majority rule*?

Answers

1. (a) A candidate would need all 24,000 votes to win a unanimous victory.
   (b) A candidate would need at least 12,001 votes to win a majority victory.
   (c) It is possible for a candidate to win a plurality victory with any number of votes more than $24,000 \div 4 = 6,000$. To win with 6,001, however, would require two of the others to get 6,000 and one to get 5,999. The smallest number of votes to guarantee victory would be a majority (more than 12,000).

2. (a) If the 100,000 votes were split equally, there would be an eight-way tie with each candidate winning 12,500 votes. At this point, one vote would make the difference. It is thus possible for a candidate to win with a 12,501-vote plurality. Six of the other candidates would have to receive 12,500 votes, while one candidate received 12,499.
   (b) 12,501 votes out of 100,000 is approximately 12.5% of the vote.
   (c) A candidate could win all 100,000 votes, making the election unanimous.

3. A majority candidate is one with over half of the votes. Each of the other candidates must thus have less than half of the votes. This means that the majority candidate has more votes than anyone else, and thus will win the election if the plurality voting method is used. It follows that the plurality voting method is always fair by the majority fairness criterion.

4. Many elections in the United States have only two candidates, and thus the winner is chosen by a majority of the voters. However, it is more correct to call the voting *plurality rule* because of the elections with three or more voters where there is no majority candidate.

Preference Schedules and Instant Runoff Voting

One problem with plurality voting is that a small percentage of the voters can decide for everyone when there are many candidates. Another problem is that plurality voting does not consider people's second and third choices.

For example, suppose a theatre group votes to decide whether to next perform *Death of a Salesman* (D), *Equus* (E), *Hamlet* (H), or *A Raisin in the Sun* (R). The votes are given in Table 12-2.

TABLE 12-2

Votes	3	2	2	1
Play	H	D	R	E

The plurality voting method will choose *Hamlet*, even though only $\frac{3}{8} = 37.5\%$ of the group voted for it. Furthermore, the other 62.5% wanted a modern play, but a classic was chosen instead.

To better understand how each voter feels about the four choices, we could ask them to list out the four plays in their order of preference. The result might be Table 12-3.

TABLE 12-3

First choice	H	H	H	D	D	R	R	E
Second choice	D	D	R	E	E	D	E	D
Third choice	R	R	D	R	R	E	D	H
Last choice	E	E	E	H	H	H	H	R

By combining identical votes, we arrive at the *preference schedule* shown in Table 12-4.

TABLE 12-4

Number of votes	2	1	2	1	1	1
First choice	H	H	D	R	R	E
Second choice	D	R	E	D	E	D
Third choice	R	D	R	E	D	H
Last choice	E	E	H	H	H	R

The *instant runoff voting method* decides elections as follows:

1. Total the number of first-choice votes for each candidate.
2. Eliminate the candidate(s) with the fewest first-choice votes.
3. Repeat the process (hold an immediate runoff election) using the top not-yet-eliminated candidate for each voter.
4. The winner is the last candidate to be eliminated.

In Table 12-4, for example, *Equus* has only one first-choice vote, while all the others have two or more. Thus, E is eliminated from the preference schedule, resulting in Table 12-5.

TABLE 12-5

Number of votes	2	1	2	1	1	1
First choice	H	H	D	R	R	
Second choice	D	R		D		D
Third choice	R	D	R		D	H
Last choice			H	H	H	R

We now repeat the process as if the last voter had voted D as top choice, instead of the previous E. This means that H and D each have three first-choice votes, while R has only two. We thus eliminate choice R, above, resulting in Table 12-6.

TABLE 12-6

Number of votes	2	1	2	1	1	1
First choice	H	H	D			
Second choice	D			D		D
Third choice		D			D	H
Last choice			H	H	H	

Here, D has five top-choice votes, while H has only three. Thus, *Death of a Salesman* will win the election if the instant runoff voting method is used.

SOLVED PROBLEMS

Preference Schedules and Instant Runoff Voting

1. Suppose the 14 members of a softball team want to decide whether to go out for pizza (P), burgers (B), or Mexican food (M). When each player writes out these three options in order of preference, the results are MPB, PBM, BPM, BPM, PBM, BMP, MPB, PBM, MPB, MPB, BMP, BPM, MBP, and MBP.

   (a) Write this out as a preference schedule.
   (b) Which choice will win if plurality voting is used?
   (c) Which choice will win if instant runoff voting is used?

2. Suppose an election is held to decide among choices A, B, C, and D, resulting in the preference schedule in Table 12-7.

   TABLE 12-7

Number of votes	3	3	5	3	2	3	2
First choice	A	A	B	C	C	D	C
Second choice	C	B	C	A	B	B	D
Third choice	B	C	D	B	A	A	A
Last choice	D	D	A	D	D	C	B

   (a) How many people voted?
   (b) How many people picked A as their first choice?
   (c) Is there a majority choice?
   (d) Which choice will the plurality method pick?
   (e) Which choice will the instant runoff method pick?

3. Will elections determined by instant runoff voting always be fair by the majority fairness criterion?

   Answers

   1. (a)

Number of votes	4	3	3	2	2
First choice	M	P	B	B	M
Second choice	P	B	P	M	B
Last choice	B	M	M	P	P

   (b) Mexican food has a plurality with six first-choice votes, compared with three for pizza and five for burgers. Thus, Mexican food will win if the plurality voting method is used.

   (c) If instant runoff voting is used, pizza will be eliminated first because it has the fewest first-choice votes. The three people who picked pizza first will now have their votes go to their second choice: burgers. This means that burgers will beat Mexican food with eight top-choice votes to six. The end result is that burgers will win if the instant runoff voting method is used.

   2. (a) A total of 21 people voted.
      (b) Choice A received six first-choice votes.
      (c) A majority choice would require at least 11 of the 21 first-place votes. None of the choices has this many first-place votes.

(d) The plurality voting method will pick choice C because it has seven top-place votes, more than any other choice.

(e) If instant runoff voting is used, D will be eliminated first, then A. Between just B and C, 11 voters put B higher on their lists and ten put C higher. Thus, the instant runoff voting method will pick choice B.

3. A majority candidate will always have more top-choice votes than any other candidate and thus will never be eliminated. This means that a majority candidate, if there is one, will always win elections determined by instant runoff voting. Thus, instant runoff voting is always fair by the majority fairness criterion.

Pairwise Comparisons and the Condorcet Criterion

The *pairwise comparison voting method* works by looking at only two candidates at a time. A candidate wins one point for each opponent beaten in such one-on-one matches. In the event of a tie, each candidate is awarded half a point. The candidate with the most points wins.

For example, suppose the ten employees of an office vote their radio preferences among classic rock (R), easy listening (E), and no radio at all (N), as shown in Table 12-8.

TABLE 12-8

Number of votes	1	4	2	3
First choice	N	R	N	E
Second choice	R	N	E	R
Last choice	E	E	R	N

If it were just a matter of easy listening or classic rock, there would be a 5–5 tie, as shown in Table 12-9.

TABLE 12-9

Number of votes	1	4	2	3
First choice		R		E
Second choice	R		E	R
Last choice	E	E	R	

If the choice of no music were compared one-on-one with classic rock, then rock would win by 7 to 3, as shown in Table 12-10.

TABLE 12-10

Number of votes	1	4	2	3
First choice	N	R	N	
Second choice	R	N		R
Last choice			R	N

Finally, the option of no music beats easy listening by 7–3 if matched up, as in Table 12-11.

Using the pairwise comparison voting method, R wins 1.5 points: one for beating N and 0.5 for tying with E. Option N wins one point for beating E. Option E wins 0.5 points for tying with R. Thus, classic rock wins the election.

TABLE 12-11

Number of votes	1	4	2	3
First choice	N		N	E
Second choice		N	E	
Last choice	E	E		N

A candidate who beats all the others (no ties) in pairwise comparisons is called a *Condorcet candidate*. Many elections, like the one in Table 12-8, do not have a Condorcet candidate. If there is a Condorcet candidate, the *Condorcet fairness criterion* states that it would be unfair if a voting method awarded victory to anyone else.

SOLVED PROBLEMS

Pairwise Comparisons and the Condorcet Criterion

1. Given the preference schedule given in Table 12-12, which candidate will win by the pairwise comparison voting method?

TABLE 12-12

Number of votes	5	4	3	2	1	3
First choice	A	D	C	D	C	E
Second choice	B	A	B	C	E	B
Third choice	C	B	E	A	A	C
Fourth choice	D	C	D	B	B	D
Last choice	E	E	A	E	D	A

2. Suppose Table 12-13 is the preference schedule for an election.

TABLE 12-13

Number of votes	1	2	2	3
First choice	A	B	C	D
Second choice	B	A	A	A
Third choice	C	C	D	B
Last choice	D	D	B	C

(a) Which candidate will win via the plurality voting method?
(b) Is there a majority candidate?
(c) Which candidate will win via the instant runoff voting method?
(d) Is there a Condorcet candidate?
(e) Which candidate will win via the pairwise comparison voting method?
(f) What conclusions about voting method fairness can be drawn from this election?

3. Why is a majority candidate always a Condorcet candidate?
4. Are all Condorcet candidates majority candidates?
5. Is the pairwise comparison voting method always fair by the Condorcet criterion? Is the pairwise comparison voting method always fair by the majority criterion?

6. When using the pairwise comparison voting method, how many comparisons must be made when there are (a) three, (b) four, (c) five, (d) six, and (e) ten candidates?

Answers

1. A beats B by 12–6. A ties with C. A loses to D by 6–12. A beats E by 11–7. B beats C by 12–6. B beats D by 12–6. B beats E by 14–4. C beats D by 12–6. C beats E by 15–3. D beats E by 11–7. The end result is that A gets 2.5 points, B gets three points, C gets 2.5 points, D gets two points, and E gets no points. Thus, when the pairwise comparison voting method is used for this election, the winner is B.

2. (a) D will win with three first-place votes when the plurality voting method is used.
   (b) There is no candidate with the necessary five of the eight first-place votes needed for a majority.
   (c) When the instant runoff method is used, first A is eliminated, then C, and then B, so D wins.
   (d) Candidate A beats B, C, and D in one-on-one comparisons; thus, A is a Condorcet candidate.
   (e) Using the pairwise comparison method, A earns three points for beating B, C, and D. Candidate B earns one point for beating C, C earns one point for beating D, and D earns one point for beating B. Thus, A wins the election when the pairwise comparison voting method is used.
   (f) A is a Condorcet candidate, but D wins when either the plurality or instant runoff voting method is used. This means that neither the plurality nor instant runoff voting method is always fair by the Condorcet criterion.

3. A majority candidate has more than half of the first-place votes, and thus will defeat every other candidate in one-on-one comparisons. Thus, every majority candidate is a Condorcet candidate.

4. It is possible to be a Condorcet candidate and not a majority candidate, as illustrated by candidate A in exercise (2).

5. A Condorcet candidate wins every one-on-one comparison, and thus will earn the maximum possible number of points when the pairwise comparison voting method is used. No other candidate can earn as many because they will lose to the Condorcet candidate. It follows that the pairwise comparison voting method is always fair by the Condorcet criterion.

   A majority candidate is automatically a Condorcet candidate, and thus will always win when the pairwise comparison voting method is used. Thus, the pairwise comparison voting method will always be fair by the majority criterion.

6. The number of comparisons among n candidates is the number of ways two different candidates can be chosen out of the n, which is defined combinatorially as $\binom{n}{2} = \dfrac{n!}{2!(n-2)!} = \dfrac{n \cdot (n-1)}{2}$. Thus:

   (a) $\binom{3}{2} = \dfrac{3 \cdot 2}{2} = 3$ comparisons

   (b) $\binom{4}{2} = \dfrac{4 \cdot 3}{2} = 6$ comparisons

   (c) $\binom{5}{2} = \dfrac{5 \cdot 4}{2} = 10$ comparisons

   (d) $\binom{6}{2} = \dfrac{6 \cdot 5}{2} = 15$ comparisons

   (e) $\binom{10}{2} = \dfrac{10 \cdot 9}{2} = 45$ comparisons

The Borda Count Voting Method

The *Borda count voting method* calculates the winner of an election by assigning points for every vote cast:

1. One point is given for every last-place vote a candidate receives.
2. Two points are given for every second-to-last-place vote a candidate receives.
3. Three points are given for every third-to-last-place vote a candidate receives, etc.
4. The candidate with the most points wins.

For example, suppose ten kids vote their preferences for going to a park (P), to a zoo (Z), to swim (S), or on a hike (H), as shown in Table 12-14.

TABLE 12-14

Number of votes	3	2	2	1	1	1
First choice	Z	H	P	S	S	H
Second choice	S	S	S	P	H	P
Third choice	H	P	H	H	P	S
Last choice	P	Z	Z	Z	Z	Z

The Borda count voting method assigns 23 points to option P, as calculated in Table 12-15.

TABLE 12-15

NUMBER OF VOTES	3	2	2	1	1	1	TOTAL	VALUE	POINTS EARNED	TOTAL POINTS
First choice			P				2	4 points each	$2 \times 4 = 8$	
Second choice				P		P	2	3 points each	$2 \times 3 = 6$	
Third choice		P			P		3	2 points each	$3 \times 2 = 6$	
Last choice	P						3	1 point each	$3 \times 1 = 3$	$8 + 6 + 6 + 3 = 23$

Option Z earns 19 points, as calculated in Table 12-16.

TABLE 12-16

NUMBER OF VOTES	3	2	2	1	1	1	TOTAL	VALUE	POINTS EARNED	TOTAL POINTS
First choice	Z						3	4 points each	$3 \times 4 = 12$	
Second choice							0	3 points each	$0 \times 3 = 0$	
Third choice							0	2 points each	$0 \times 2 = 0$	
Last choice		Z	Z	Z	Z	Z	7	1 point each	$7 \times 1 = 7$	$12 + 7 = 19$

Option H earns 27 points, as calculated in Table 12-17.

TABLE 12-17

NUMBER OF VOTES	3	2	2	1	1	1	TOTAL	VALUE	POINTS EARNED	TOTAL POINTS
First choice		H				H	3	4 points each	$3 \times 4 = 12$	
Second choice				H			1	3 points each	$1 \times 3 = 3$	
Third choice	H		H	H			6	2 points each	$6 \times 2 = 12$	
Last choice							0	1 point each	$0 \times 1 = 0$	$12 + 3 + 12 = 27$

Option S earns 31 points, as calculated in Table 12-18.

TABLE 12-18

NUMBER OF VOTES	3	2	2	1	1	1	TOTAL	VALUE	POINTS EARNED	TOTAL POINTS
First choice				S	S		2	4 points each	$2 \times 4 = 8$	
Second choice	S	S	S				7	3 points each	$7 \times 3 = 21$	
Third choice						S	1	2 points each	$1 \times 2 = 2$	
Last choice							0	1 point each	$0 \times 1 = 0$	$8 + 21 + 2 = 31$

Because option S earns the most points, it wins the election when the Borda count voting method is used.

There are many possible variations on this method. Zero points might be awarded for last-place votes, for example, instead of one. Similarly, an extra point or more could be given for first-place votes. Each variation results in a new voting method. The point values for votes under the Borda count method are merely the easiest to explain.

SOLVED PROBLEMS

The Borda Count Voting Method

1. Suppose an election results in the preference schedule given in Table 12-19. Find the winner when the Borda count voting method is used.

TABLE 12-19

Number of votes	3	1	2	2	4	1
First choice	A	A	B	B	C	C
Second choice	B	C	A	C	A	B
Third choice	C	B	C	A	B	A

2. Suppose an election with five candidates results in the preference schedule shown in Table 12-20.

TABLE 12-20

Number of votes	2	3	6
First choice	A	B	C
Second choice	B	D	A
Third choice	D	A	B
Fourth choice	E	E	E
Last choice	C	C	D

(a) Which candidate will win if the Borda count voting method is used?

(b) Is there a majority candidate?

(c) Is there a Condorcet candidate?

(d) What does this election illustrate?

3. What would happen if ten points were given for each first-choice vote and no points were given for anything else?

Answers

1. With three candidates, the Borda count voting method assigns one point for each last-place vote, two points for each second-place vote, and three points for each first-choice vote.

    Candidate A has three last-place votes, six second-choice votes, and four first-place votes. Thus, A receives
    $(3 \times 1) + (6 \times 2) + (4 \times 3) = 27$ points.

    Candidate B has five last-place votes, four second-place votes, and four first-place votes. Thus, B receives
    $(5 \times 1) + (4 \times 2) + (4 \times 3) = 25$ points.

    Candidate C has five last-place votes, three second-place votes, and five first-place votes. Thus, C receives
    $(5 \times 1) + (3 \times 2) + (5 \times 3) = 26$ points.

    If the Borda count voting method is used, candidate A will win the election.

2. Candidate A has no last-place votes (one point each), no fourth-choice votes (two points each), three third-choice votes (three points each), six second-choice votes (four points each), and two top-choice votes (five points each). This gives A a total of $(3 \times 3) + (6 \times 4) + (2 \times 5) = 43$ points.

    Candidate B has six third-place votes, two second-place votes, and three top-choice votes for a total of $(6 \times 3) + (2 \times 4) + (3 \times 5) = 41$ points.

    Candidate C has five last-place votes and six first-place votes, for a total of $(5 \times 1) + (6 \times 5) = 35$ points.

    Candidate D has six last-place votes, two third-place votes, and three second-place votes, for a total of $(6 \times 1) + (2 \times 3) + (3 \times 4) = 24$ points.

    Candidate E has 11 second-to-last-place votes, for a total of $11 \times 2 = 22$ points.

    (a) When the Borda count voting method is used, A wins this election.
    (b) C is a majority candidate, with six of 11 first-place votes.
    (c) Because majority candidates are automatically Condorcet candidates, C is also a Condorcet candidate.
    (d) Here, candidate C is both a majority and Condorcet candidate, but does not win the election when the Borda count voting method is used. This illustrates that the Borda count voting method is not always fair by the majority and Condorcet fairness criteria.

3. A voting system that only awards points to the first-choice votes will be the same as the plurality voting method. Both methods only consider each voter's top choice.

Two More Fairness Criteria

With four different voting methods, it will help to have more fairness criteria with which to compare them.

The *monotonicity fairness criterion* states that it would be unfair if increasing the number of votes for the winner were to make him or her lose.

For example, suppose candidate A were about to win an election. Then, at the last moment, some supporters of other candidates changed their votes to make candidate A top choice. If the plurality voting method were used, this would only increase the margin of A's victory. If the Borda count voting method were used, this would increase the number of points for A and decrease the points of other candidates. If the pairwise comparison voting method were used, A would win the same matches as before, perhaps more. With all these voting methods, candidate A would still win the election. Thus, the plurality, Borda count, and pairwise comparison voting methods are always fair by the monotonicity fairness criterion.

The instant runoff voting method is not always fair by the monotonicity fairness criterion. For example, suppose a group of 21 people have decided to hold an election using the instant runoff voting method. On the eve of the election, it is well-known how each voter feels, as illustrated by the preference schedule in Table 12-21.

TABLE 12-21

Number of votes	5	6	8	2
First choice	A	B	C	A
Second choice	B	C	A	C
Third choice	C	A	B	B

With this schedule, B will be eliminated first and then A, so C will win the election. Now suppose that the last two voters, the ones who voted A-C-B, decide to change their votes to C-A-B. This only increases the number of first-place votes for C. By the monotonicity fairness criterion, this ought not change the outcome of the election. However, the new preference schedule is illustrated in Table 12-22.

TABLE 12-22

Number of votes	5	6	8	2
First choice	A	B	C	C
Second choice	B	C	A	A
Third choice	C	A	B	B

Now when the instant runoff voting method is applied, A will be eliminated first. Candidate B will then beat C by 11–10, thus winning the election.

Increasing the number of first-place votes for C has caused C to lose the election. This illustrates how the instant runoff voting method can be unfair by the monotonicity fairness criterion.

Another fairness criterion is the *independence-of-irrelevant-outcomes fairness criterion*. This states that it would be unfair if the outcome of an election were changed by removing an option that did not win.

For example, suppose a company decided to move to Chicago. If they were then told that they could not move to London, it should not change their decision. As obvious as this may sound, all four of our voting methods can violate this fairness criterion.

In the American plurality voting system, any votes for a disqualified candidate are thrown away. This will not change the outcome of the election. In this sense, the voting system is fair by the independence-of-irrelevant-outcomes fairness criterion. However, if the supporters of the rejected candidate are asked their second choices, the outcome can change.

For example, suppose that when Bill, George, and Ross run for office, the resulting preference schedule is given in Table 12-23.

TABLE 12-23

Percentage of all votes	43%	38%	19%
First choice	Bill	George	Ross
Second choice	George	Bill	George
Third choice	Ross	Ross	Bill

Using the plurality voting system, Bill will win the election. George and Ross lose. If Ross were not on the ballot, his supporters would vote for George instead. This would give George a 57% majority of the vote. Thus, removing a losing option can change the outcome of the election. This example proves that the plurality voting method can violate the independence-of-irrelevant-outcomes fairness criterion.

SOLVED PROBLEMS

Two More Fairness Criteria

1. Suppose an election using the instant runoff voting method results in the preference schedule given in Table 12-24.

   (a) Who will win?
   (b) Who will lose?
   (c) If candidate B drops out, what will the preference schedule look like?
   (d) Who will win the election if B drops out?

TABLE 12-24

Number of votes	3	2	2	1
First choice	A	B	C	D
Second choice	D	D	D	C
Third choice	B	C	B	B
Last choice	C	A	A	A

   (e) What does this election illustrate?

2. Suppose an election using the Borda count voting method results in the preference schedule given in Table 12-25.

   (a) Who will win the election?
   (b) Who will lose the election?
   (c) If candidate C drops out, what will the preference schedule look like?

TABLE 12-25

Number of votes	2	4	7
First choice	A	D	B
Second choice	B	A	D
Third choice	C	C	C
Fourth choice	D	B	A

   (d) Who will win the election if C drops out?
   (e) What does this election illustrate?

3. Suppose an election using the pairwise comparison voting method results in the preference schedule given in Table 12-26.

   (a) Who will win this election?
   (b) Who will lose this election?

TABLE 12-26

Number of votes	4	2	6	4	4	1	1
First choice	A	B	C	C	D	E	E
Second choice	B	A	B	B	E	A	C
Third choice	D	E	E	A	A	B	B
Fourth choice	E	C	A	D	C	C	A
Last choice	C	D	D	E	B	D	D

(c) What will the preference schedule look like if candidate E drops out?

(d) Who will win if candidate E drops out?

(e) What does this election illustrate?

Answers

1.   (a)   First D will be eliminated, then B, and then A, so C will win.

      (b)   Candidates A, B, and D will all lose.

      (c)   If B drops out, the new preference schedule is given in Table 12-27.

TABLE 12-27

Number of votes	3	2	2	1
First choice	A	D	C	D
Second choice	D	C	D	C
Last choice	C	A	A	A

      (d)   C will be eliminated first, and then A, so D will win the election.

      (e)   Removing a loser from the ballot has changed the outcome of the election. This election illustrates how the instant runoff voting method can be unfair by the independence-of-irrelevant-outcomes fairness criterion.

2.   (a)   A will receive $(7 \times 1) + (4 \times 3) + (2 \times 4) = 27$ points. B will receive $(4 \times 1) + (2 \times 3) + (7 \times 4) = 38$ points. C will receive $13 \times 2 = 26$ points. D will receive $(2 \times 1) + (7 \times 3) + (4 \times 4) = 39$ points. Thus, D will win the election when the Borda count voting method is used.

      (b)   Because D wins, A, B, and C will all lose.

      (c)   If C drops out of the election, the result will be the preference schedule in Table 12-28.

TABLE 12-28

Number of votes	2	4	7
First choice	A	D	B
Second choice	B	A	D
Last choice	D	B	A

      (d)   Now A will receive $(7 \times 1) + (4 \times 2) + (2 \times 3) = 21$ points. B will receive $(4 \times 1) + (2 \times 2) + (7 \times 3) = 29$ points. D will receive $(2 \times 1) + (7 \times 2) + (4 \times 3) = 28$ points. Thus, B will win the election.

      (e)   This election illustrates how the Borda count voting method can be unfair by the independence-of-irrelevant-outcomes fairness criterion. Removing an irrelevant outcome (choice C) has changed the outcome of the election.

3. (a) A beats D and ties with C, so A earns 1.5 points. B beats A, D, and E, so B earns three points. C beats B, beats D, and ties with A, so C earns 2.5 points. D beats E, earning 1 point. E beats A and C, so E earns 2 points. Thus, B wins the election.

   (b) Because B wins, A, C, D, and E all lose.

   (c) If candidate E drops out, the result will be the preference schedule given in Table 12-29.

TABLE 12-29

Number of votes	4	2	6	4	4	1	1
First choice	A	B	C	C	D	A	C
Second choice	B	A	B	B	A	B	B
Third choice	D	C	A	A	C	C	A
Last choice	C	D	D	D	B	D	D

   (d) A beats D and ties with C, so A earns 1.5 points. B beats A and D, earning two points. C beats B, beats D, and ties with A, so C earns 2.5 points. D beats no other candidate, so D earns 0 points. Thus, C wins the election.

   (e) The outcome of this election depends on whether candidate E stays in the election or drops out, even though E will not win. This illustrates how the pairwise comparison voting method is not always fair by the independence-of-irrelevant-outcomes fairness criterion.

Arrow's Impossibility Theorem

Every one of the voting methods described so far has been shown capable of violating one of the fairness criteria. This is not unusual. In 1949, Kenneth Arrow proved that every conceivable voting method will occasionally fail one of the fairness criteria. No matter how the votes are counted, it will be possible to set up a preference schedule for which the voting method fails one of the fairness criteria. This is known as *Arrow's impossibility theorem*.

SUPPLEMENTAL PROBLEMS

1. A large family casts votes to decide what to watch on TV. The choices are *Toolboxing* (T); *Lives of the Serfs* (S); *Janelle, Model Detective* (J); and *Lawsuit!* (L). When each person lists his or her preferences, the results are STLJ, JTLS, TLJS, SLTJ, JLST, SLTJ, TLJS, JTLS, and SLTJ.

   (a) Write the election results as a preference schedule.
   (b) Is there a majority candidate?
   (c) Which show will win if the plurality voting system is used?
   (d) Which show will win if instant runoff voting is used?
   (e) Which show will win if the Borda count voting method is used?

2. The Dance Collective is in dire need of money. The only fundraising ideas they have are a bake sale (B), a raffle (R), to apply for a grant (G), or to appeal to charity (C) by begging. When everyone has voted by listing their options in order, the result is the preference schedule in Table 12-30.

   (a) How many members voted?
   (b) What percentage of the voters think a grant is the best option?
   (c) Which option will win if plurality voting is used?
   (d) Which option will win if instant runoff voting is used?

TABLE 12-30

Number of votes	8	4	9	6	3	3
First choice	G	B	R	C	B	G
Second choice	R	R	B	R	G	C
Third choice	B	C	G	G	R	R
Last choice	C	G	C	B	C	B

(e) Which option will win if the pairwise comparison voting method is used?

(f) Which option will win via the Borda count voting method?

3. The eight members of a committee must decide a time for their weekly meetings. They have reduced the options to (A) Tuesdays at 8 A.M., (B) Tuesdays at 11 A.M., (C) Wednesdays at 4 P.M., and (D) Fridays at 8 A.M. When they list their preferences, the end result is Table 12-31.

TABLE 12-31

Number of votes	3	2	1	2
First choice	D	A	B	C
Second choice	A	B	A	B
Third choice	B	D	C	A
Last choice	C	C	D	D

(a) Does any option have a majority?

(b) Which option has a plurality of the votes?

(c) Is there a Condorcet choice?

(d) What sort of unfairness is illustrated here?

4. At a secret ninja convention, 32% declare "death from above" to be their favorite move. There are 38% who favor "death from behind." The remaining 30% like "throwing stars from a distance" best. When further pressed, the preference schedule in Table 12-32 is obtained.

TABLE 12-32

Votes	32%	18%	30%	20%
First choice	Above	Behind	Stars	Behind
Second choice	Stars	Above	Behind	Stars
Last choice	Behind	Stars	Above	Above

What is the ultimate (favorite) ninja move when judged by the (a) plurality, (b) instant runoff, (c) pairwise comparison, or (d) Borda count voting method?

5. Find the winner of the election illustrated in Table 12-33 via the (a) plurality, (b) instant runoff, (c) pairwise comparison, and (d) Borda count voting method.

TABLE 12-33

Number of votes	3	3	5	4	2	3
Top choice	A	D	D	B	E	B
Second choice	C	B	C	C	A	E
Third choice	E	C	E	A	C	C
Fourth choice	B	A	A	D	B	A
Last choice	D	E	B	E	D	D

6. A poll at the annual Big Bad Wolf Symposium results in the preference schedule for Preferred Pig House Construction Materials given in Table 12-34. What sort of election is this?

TABLE 12-34

Votes	100%
First choice	Straw
Second choice	Wood
Last choice	Bricks

7. Explain why it is crucial for a group to decide upon one voting method before the votes in an election are cast.
8. Complete Table 12-35 by writing "fair" when the given voting method is always fair by the given fairness criterion, and write "unfair" otherwise.

TABLE 12-35

	PLURALITY	INSTANT RUNOFF	PAIRWISE COMPARISON	BORDA COUNT
Majority				
Condorcet				
Monotonicity				
Independence of irrelevant outcomes				

9. Suppose an organization feels that the majority criterion is the most important of the fairness criteria. Next they value the Condorcet criterion, and then the monotonicity criterion. Least important to them is the independence-of-irrelevant-outcomes criterion. Which voting method would be best for them to use?
10. Suppose elections are evaluated by the pairwise comparison voting method. How many comparisons will be necessary if there are (a) eight, (b) 12, (c) 15, and (d) 20 options?
11. Suppose a man at a restaurant decides to order tomato soup with his meal. Before he orders, the waiter informs him that they are all out of clam chowder. Because of this, the man changes his mind and orders chicken soup instead. What fairness criterion does this example violate?
12. Could a Condorcet candidate receive no first-place votes? Could a majority candidate receive no first-place votes? Could a plurality candidate receive no first-place votes?

Answers

1. (a) See Table 12-36.

TABLE 12-36

Number of votes	1	2	2	3	1
First choice	S	J	T	S	J
Second choice	T	T	L	L	L
Third choice	L	L	J	T	S
Fourth choice	J	S	S	J	T

   (b) No
   (c) S
   (d) J
   (e) L and T will tie

2. (a) 33
   (b) $33\frac{1}{3}\%$
   (c) G
   (d) R
   (e) R
   (f) R

3. (a) No
   (b) D
   (c) A is a Condorcet choice.
   (d) This illustrates that the plurality voting method can be unfair by the Condorcet fairness criterion.

4. (a) "Death from behind"
   (b) "Death from behind"
   (c) "Throwing stars from a distance"
   (d) "Throwing stars from a distance"

5. (a) D
   (b) B
   (c) C
   (d) C

6. Unanimous
7. Different voting methods will often choose different winners from the same voting schedule. To avoid conflict, it is thus best to agree upon the voting method before casting votes.
8. See Table 12-37.

TABLE 12-37

	PLURALITY	INSTANT RUNOFF	PAIRWISE COMPARISON	BORDA COUNT
Majority	Fair	Fair	Fair	Unfair
Condorcet	Unfair	Unfair	Fair	Unfair
Monotonicity	Fair	Unfair	Fair	Fair
Independence of irrelevant outcomes	Unfair	Unfair	Unfair	Unfair

9. Pairwise comparison is the only voting method which always is fair by their top three fairness criteria.
10. (a) 28, (b) 66, (c) 105, and (d) 190
11. This decision violates the independence-of irrelevant-outcomes-fairness criterion.
12. It is possible for a Condorcet candidate to receive no first-place votes. However, a candidate with no first-place votes could be neither a majority nor a plurality candidate.

CHAPTER 13

Transformations and Symmetry

A *distance-preserving transformation* is a way to move an object without changing the object's size or shape. In this chapter, we shall use the word *transformation* to refer to such moves. The end result of a transformation is called the *image* of the object under the transformation. Transformations are also called *rigid motions* because we imagine the object is firm enough to hold its shape when moved. It is easier to move a shape cut out of cardboard, for example, than it is to move a design made of toothpicks.

The *symmetries* of an object are the transformations that leave it in the same place and looking the same.

Translations

The most basic transformation is the slide, usually called a *translation*. When an object slides left, right, up, or down, its shape and size remain unchanged.

As an example, one could translate an object by sliding it 5 units to the right and 2 units up. If the object was the triangle with vertices (0, 0), (1, 2), and (4, −1), then its image under this transformation would be the triangle with vertices (5, 2), (6, 4), and (9, 2), as shown in Fig. 13-1.

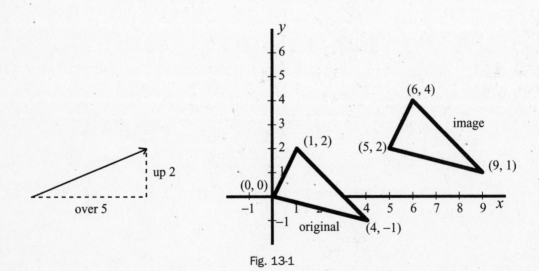

Fig. 13-1

SOLVED PROBLEMS

Translations

1. Draw the image of the triangle in Fig. 13-2 under the translation that

   (a) slides everything down 3 units.
   (b) slides everything to the right 4 units.
   (c) slides everything down 1 unit and to the left 3 units.
   (d) slides everything to the right 6 units and up 2 units.

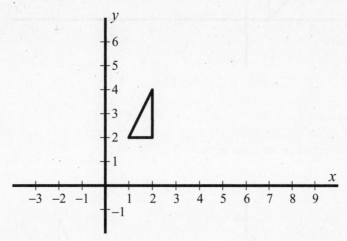

Fig. 13-2

2. Which of the shapes in Fig. 13-3 could be the result of translating rectangle 1? If possible, name the translation.

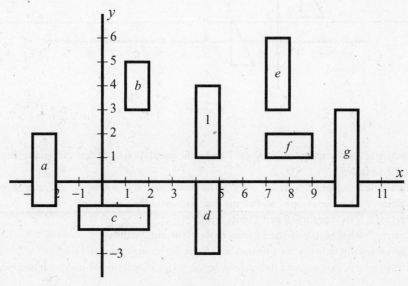

Fig. 13-3

3. Translate the triangle in Fig. 13-4 by sliding it to the right 4 units and down 1 unit. Draw the image. Next, translate this new triangle to the right 2 units and up 3 units. Draw the final result. Is the third triangle a translation of the first?

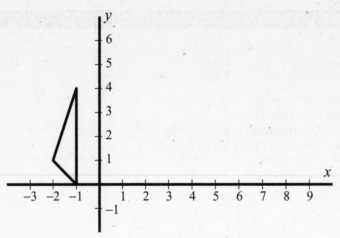

Fig. 13-4

Answers

1. The images of the triangle are given in Fig. 13-5.

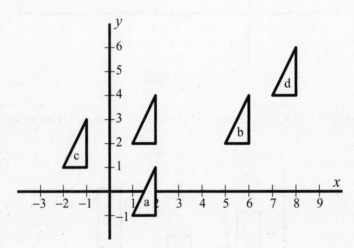

Fig. 13-5

2. (a) Rectangle *a* is the image of rectangle 1 under the translation that takes everything 7 units to the left and 2 units down.
    (b) Rectangle *b* cannot be the result of a translation because it is not the same size or shape as rectangle 1.
    (c) Rectangle *c* has been turned, something that no translation can do.
    (d) Rectangle *d* has been translated 4 units down.
    (e) Rectangle *e* has been translated 3 units to the right and 2 units up.
    (f) Rectangle *f* not only has been rotated but also is the wrong shape and size.
    (g) Rectangle *g* is too long, and thus cannot be the image of rectangle 1 under a translation.

3. The results of the two translations are illustrated in Fig. 13-6. The third triangle is the image of the first under the translation that shifts everything to the right 6 units and up 2 units. Just as in this example, the combination of any two translations will be another translation.

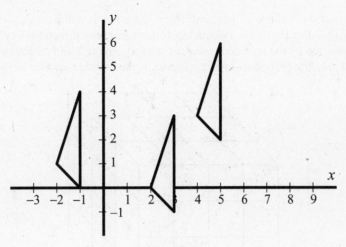

Fig. 13-6

Translation Symmetry

An object has *translation symmetry* if it can be slid by a positive amount and end up in the same place, looking the same. The only way this could happen is if the object is infinitely long and has a repeating pattern.

For example, suppose you have a strip of wallpaper border that runs forever to the left and right, as illustrated in Fig. 13-7. If someone were to translate this strip to the right by the amount indicated by the arrow, the flowers would line up exactly. The end result would look exactly like the original, even in the same spot (there is no way that it could run further to the right than the original infinity). Thus, this object has translation symmetry.

Fig. 13-7

An object with translation symmetry in two different directions (not along the same line) will have to be a plane with regular repeating patterns. For example, the sheet of wallpaper illustrated in Fig. 13-8 has translation symmetry in the two directions indicated by arrows (if you imagine it continues on forever in all directions). Traditionally, the symmetry is indicated by the shortest positive distance in which the object can be slid to line back up with itself.

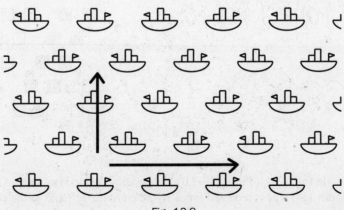

Fig. 13-8

An object with translation symmetry in three different dimensions will have patterns that repeat like the stacked-up boxes illustrated in Fig. 13-9, continuing forever in every direction. Crystals have this sort of symmetry, and are classified by the repeated patterns they form in three-dimensional space. The difference between diamond and graphite (pencil lead), for example, is the crystal structure of their carbon atoms.

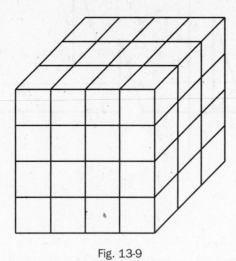

Fig. 13-9

SOLVED PROBLEMS

Translation Symmetry

For each object in Fig. 13-10, draw arrows to indicate the translation symmetry, if any.

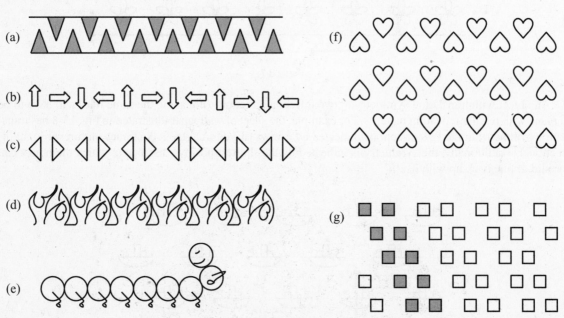

Fig. 13-10

Answers

Answers are given in Fig. 13-11. Figure 13-11(e) has no translation symmetry. Also, unless the pattern of shading in Figure 13-11(g) is repeated infinitely often to the left and to the right, there is no horizontal translation symmetry.

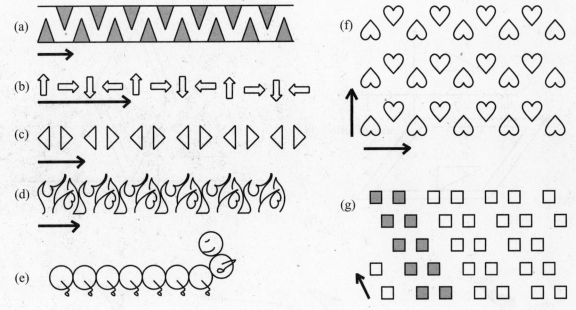

Fig. 13-11

Reflections

The next most basic transformation is the flip or *reflection*. A reflection takes an object to its mirror image.

For example, if a mirror is held to the plane with one edge along the y-axis, then the image of the triangle with vertices (−3, 1), (−1, 1) and (−1, 2) will be as illustrated in Fig. 13-12. This is called a *y-axis reflection* because of the location of the mirror. Notice that this transformation could not be the result of a translation because the triangle ends up pointing in a different direction.

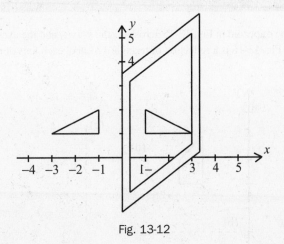

Fig. 13-12

The *axis of reflection* is the line where the edge of the mirror meets the plane. If this is changed, then the image will also be changed, as illustrated in Fig. 13-13(a) and 13-13(b).

The distance an object is from a mirror is the same as the distance its image will be from the surface of the mirror. Thus, if you draw a line perpendicular from a point to the axis of reflection, the image of the point will be an equal distance past the axis, as illustrated in Fig. 13-14. This trick can also be used to find an axis of reflection. If you connect each point to its image with a straight line, then the midpoints of these lines will define the axis.

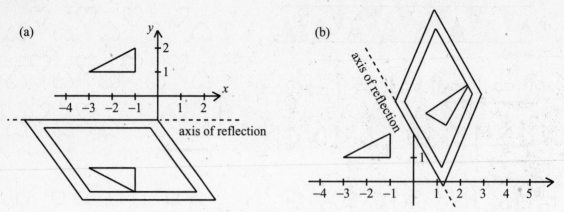

Fig. 13-13

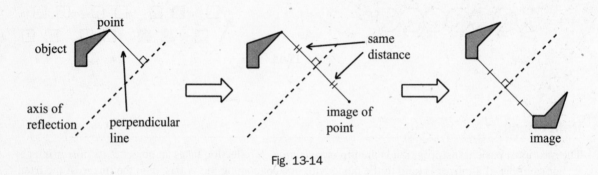

Fig. 13-14

SOLVED PROBLEMS

1. Sketch the reflection of the trapezoid in Fig. 13-15 across (a) the y-axis, and the axes labeled (b), (c), and (d).
2. Which of the triangles in Fig. 13-16 is a reflection of triangle 1? Label each axis of reflection.

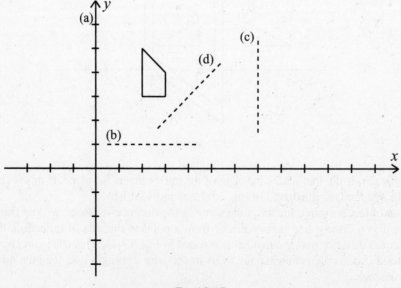

Fig. 13-15

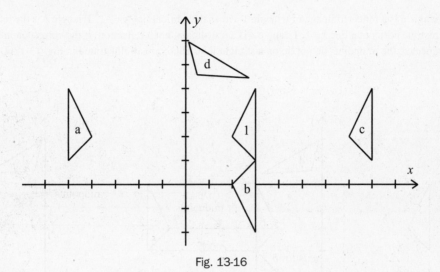

Fig. 13-16

Answers

1. The reflected images are shown in Fig. 13-17.

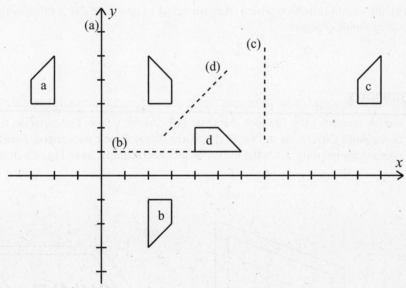

Fig. 13-17

2.   Triangle *a* is a reflected image of triangle 1 across the vertical line $y = -1$. Triangle *b* is the reflection across the horizontal line $x = 1$. Triangle *c* is a translation, not a reflection; if the corresponding points are connected, the midpoints do not lie on a straight line. These are all illustrated in Fig. 13-18(a).

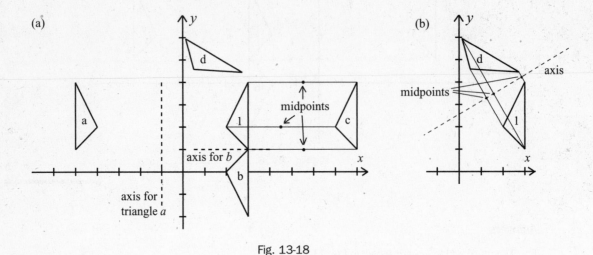

Fig. 13-18

In Fig. 13-18(b), the axis of reflection which takes triangle 1 to its image *d* is formed by the midpoints of lines connecting corresponding points.

Reflection Symmetry

An object has *reflection symmetry* if it can look the same as its mirror image. For example, when a mirror is held to the right of the word *CRATE*, as in Fig. 13-19(a), the *A* and *T* look unchanged. Similarly, the *C* and *E* have reflection symmetry, but only when the mirror is held horizontally, as in Fig. 13-19(b).

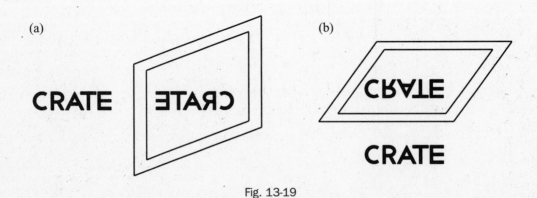

Fig. 13-19

An axis of reflection can be drawn through the center of an object with reflection symmetry so that the two halves are reflected images of each other. For example, a mirror set vertically through the center of the letter *A* will complete the letter, as illustrated in Fig. 13-20(a). This axis of reflection is called the *axis of symmetry*, usually indicated by a dotted line. The axes of symmetry for the letters *A*, *T*, *C*, and *E* are shown in Fig. 13-20(b). Some objects have several axes of symmetry. For example, the square in Fig. 13-20(c) has four different axes of symmetry.

Fig. 13-20

The reflection of an object that runs clockwise will be counterclockwise, as illustrated in Fig. 13-21(a). This can be used to recognize when objects do not have reflection symmetry. For example, the loop on the letter *R* points clockwise around the center of the letter, as shown in Fig. 13-21(b). This means that *R* does not have reflection symmetry.

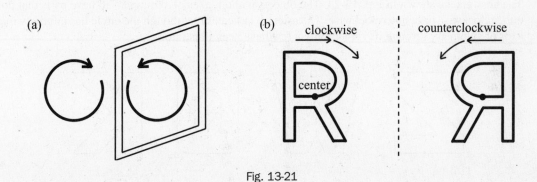

Fig. 13-21

Left and right shoes are mirror images of one another. However, though a two-dimensional object can be flipped over to become its reflected image, there is no easy way to turn a left shoe into a right one.

SOLVED PROBLEMS

Reflection Symmetry

Draw all the axes of symmetry through the objects in Fig. 13-22 that have reflection symmetry.

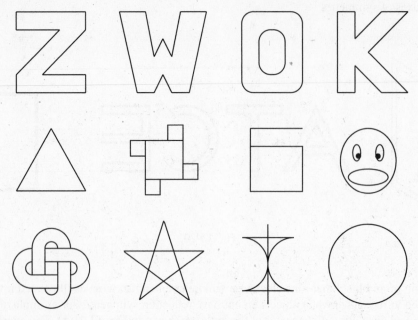

Fig. 13-22

Answers

The answers are shown in Fig. 13-23. The objects without axes of symmetry all have parts that point either clockwise or counterclockwise. The asterisk indicates that though the circle has infinitely many axes of symmetry through its center, only a few have been illustrated.

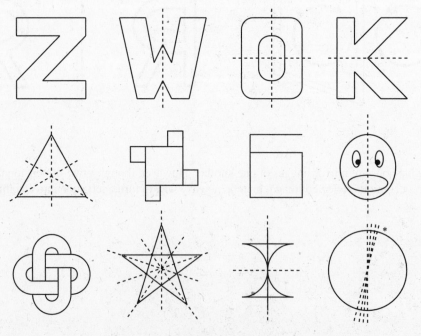

Fig. 13-23

Rotations

A *rotation* is a transformation that keeps one point fixed (called the *center of rotation*, or *rotocenter*) and moves everything else rigidly around it. Imagine a piece of paper pinned down by a single thumbtack; a rotation is a way to turn the paper around the tack.

For example, Fig. 13-24(a) illustrates a triangle rotated 90° clockwise around the origin (0, 0). A circle centered at the rotocenter that goes through a point will also go through the image of that point. The angle of rotation will be formed by the three points as illustrated in Fig. 13-24(b).

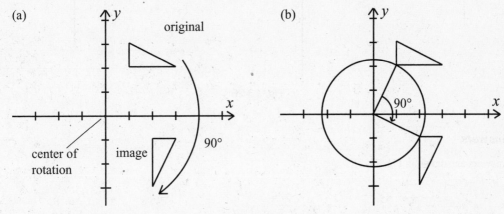

Fig. 13-24

When an object is rotated around its center, the degree of rotation is sometimes given as a fraction of a full circle. In Fig. 13-25, the letter R is rotated around its center by (a) $\frac{1}{5}$ of a full circle $\left(\frac{1}{5} \cdot 360° = 72°\right)$,

(b) $\frac{1}{4}$ of a circle (90°), (c) $\frac{1}{3}$ of a circle (120°), (d) $\frac{1}{2}$ of a circle (180°), and (e) $\frac{1}{1}$ of a circle (a full 360°). The last rotation is considered *trivial* because it does not change anything.

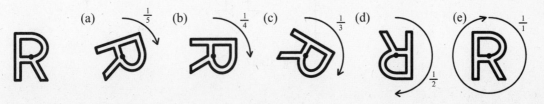

Fig. 13-25

SOLVED PROBLEMS

Rotations

1. Rotate the rectangle in Fig. 13-26(a) by 90° counterclockwise around the origin.
2. Rotate the arrow in Fig. 13-26(b) by 180° around the point (2, 1).
3. Rotate the triangle in Fig. 13-26(c) clockwise around the point (0, 2) by (a) 90°, (b) 180°, and (c) 270°.

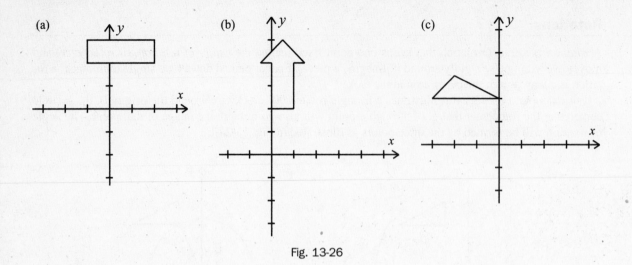

Fig. 13-26

Answers

Answers are shown in Fig. 13-27.

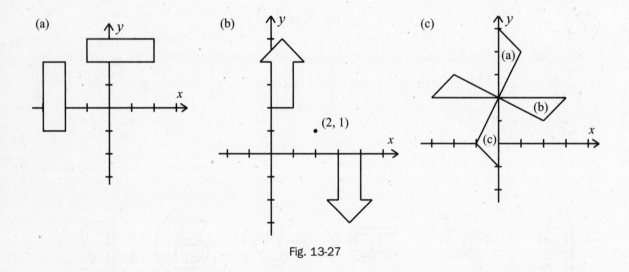

Fig. 13-27

Rotation Symmetry

If a nontrivial rotation can make an object look unchanged, then the object has *rotation symmetry*. The type of rotation is described by the number of different rotations that do this.

For example, a square can be rotated around its center by $\frac{1}{4}$ of a full circle and still look the same, as

illustrated in Fig. 13-28(a). This automatically means that it can be rotated by $\frac{1}{2}$ of a full circle (just repeat

the quarter-turn twice), by $\frac{3}{4}$ of a full circle, and by a full circle (the trivial rotation). Any smaller rotation,

for example $\frac{1}{8}$, will change the way the square looks, as shown in Fig. 13-28(b). Thus, a square has fourfold

rotation symmetry.

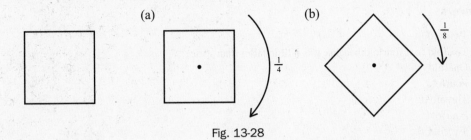

Fig. 13-28

If the letter *Z* is rotated slowly around its center, it will be halfway around before it looks like it did originally. This means that the letter *Z* has twofold symmetry; the only rotations which leave it looking the same are the half circle and the full circle.

An object with no rotation symmetry, for example the number 5, is said to have onefold symmetry. This is because every object has at least one rotation that leaves it unchanged: the trivial, complete-circle rotation.

SOLVED PROBLEMS

Rotation Symmetry

Identify the rotation symmetry of each object in Fig. 13-29.

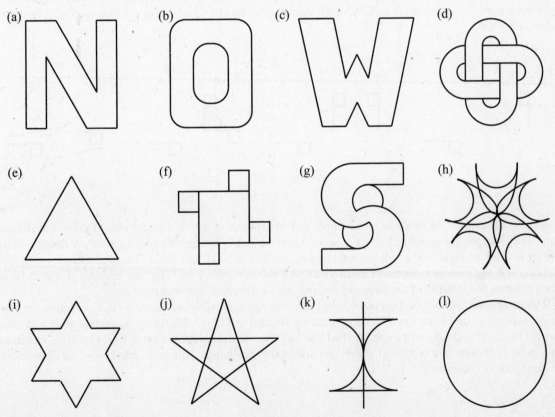

Fig. 13-29

Answers

(a) Twofold
(b) Twofold (not fourfold because it is a little taller than wide)
(c) Onefold
(d) Fourfold
(e) Threefold
(f) Fourfold
(g) Twofold
(h) Fivefold
(i) Sixfold
(j) Fivefold
(k) Twofold
(l) A circle can be rotated by any tiny amount and still look the same; thus, this special case is considered *infinite rotation symmetry*.

Combinations of Transformations

Two transformations can be combined by performing first one and then the other. Every transformation can be formed by combining rotations, reflections, and transformations. To see this, imagine any two congruent objects (same size and shape). Flip the first object over (if necessary) to match the second, then rotate it so they point in the same direction, and then slide the first to overlap the second with a translation.

For example, suppose you want a transformation that takes flag x in Fig. 13-30(a) to flag y. Because flag x waves clockwise around its center and flag y flies counterclockwise, we need to reflect x as shown in Fig. 13-30(b). Next, rotate the resulting image 90° counterclockwise, as shown in Fig. 13-30(c), so that both flags point in the same direction. Finally, this last image can be translated to exactly overlap flag y, as shown in Fig. 13-30(d).

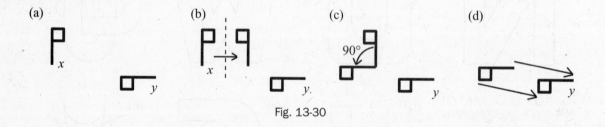

Fig. 13-30

Most combinations of rotations, reflections, and translations result in a single one of these transformations. For example, the combined effect of several translations is just another translation. Any number of rotations around a single point will be yet another rotation around that point. Two reflections result in a translation if the axes of reflection are parallel and a rotation otherwise. The combination of a translation and a rotation is a rotation of the same degree, just with a different center of rotation.

The combination of a reflection and a translation, however, is called a *glide reflection*, which is not a rotation, reflection, or translation. The transformation needed in Fig. 13-30, for example, is a glide reflection formed by the reflection given by the dashed line and the translation illustrated in Fig. 13-31(a). Traditionally, the axis of reflection is parallel to the direction of the translation. With some cleverness, this can always be arranged, as shown in Fig. 13-31(b).

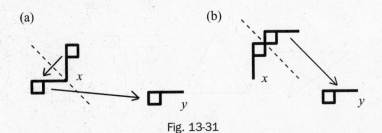

Fig. 13-31

SOLVED PROBLEMS

Combinations of Transformations

Name the sort of transformation (translation, reflection, rotation, or glide reflection) necessary to take the letter F marked 1 to each of its images in Fig. 13-32.

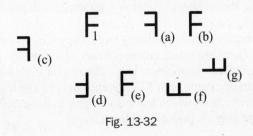

Fig. 13-32

Answers

(a) Reflection

(b) Translation

(c) This is a flipped image, but no single reflection could take 1 to (c); thus, it must be the result of a glide reflection.

(d) Rotation

(e) Translation

(f) Rotation

(g) Glide reflection

Groups of Symmetries

The collection of all the transformations which leave an object looking unchanged is called the *symmetry group*. A *group* is a set with a way to combine its elements. One element must do nothing (the trivial element). Each element must have an inverse element which does the exact opposite. The study of groups began with transformation groups but has since expanded to encompass many groups with strange and surprising properties. Symmetry groups are the natural place to begin the study of groups.

A finite object cannot remain unchanged under a translation. Thus the group of symmetries of a finite object can only contain reflections and rotations. If an object has reflection symmetry, its symmetry group is *dihedral*, represented by the letter D. If the object does not have reflection symmetry, its symmetry group is represented by the letter Z (from the Z for integers). With the letter is the number representing the rotation symmetry type.

For example, the object in Fig. 13-33(a) does not have reflection symmetry, so its symmetry group is represented by the letter Z. Because it has fourfold rotation symmetry, the symmetry group is written Z_4.

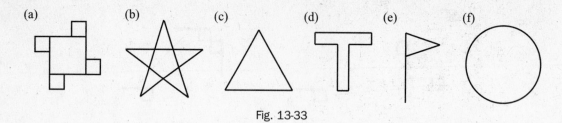

Fig. 13-33

The five-pointed star in Fig. 13-33(b) has reflection symmetry (D) and fivefold rotation symmetry, so its symmetry group is dihedral: D_5.

The triangle in Fig. 13-33(c) has reflection symmetry (D) and threefold rotation symmetry, so its symmetry group is written D_3.

The T in Fig. 13-33(d) has reflection symmetry (D), but no rotation symmetry beyond the trivial rotation. Because no rotation symmetry is considered onefold, the symmetry group of this letter is written D_1.

The flag in Fig. 13-33(e) has no reflection symmetry (Z) and no rotation symmetry (onefold). Thus, its symmetry group is Z_1.

The circle in Fig. 13-33(f) is special because it has reflection symmetry (D) and infinite rotation symmetry. Thus, its symmetry group is written D_∞.

Note that for dihedral groups, the number of rotation symmetries will always be the same as the number of axes of reflection.

Objects with translation symmetry must be infinitely large, like an infinite strip with a repeated design or an infinite sheet of wallpaper. There are seven different kinds of symmetry groups for infinite strips, depending on whether there are reflection symmetries, 180° rotations, or glide reflections which leave the strip looking the same. Wallpaper patterns come with 17 different kinds of symmetry groups, depending on the rotations, reflections, and glide reflections which leave everything looking the same. For patterns that repeat in all three dimensions, there are only six possible symmetries.

SOLVED PROBLEMS

Identify the symmetry group for each of the objects in Fig. 13-34.

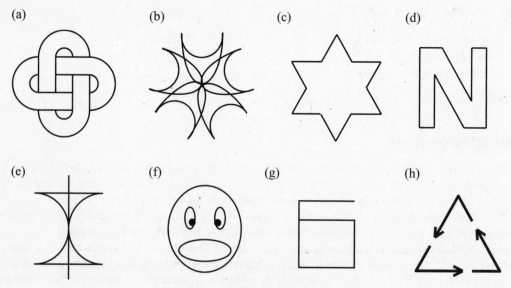

Fig. 13-34

Answers: (a) Z_4, (b) D_5, (c) D_6, (d) Z_2, (e) D_2, (f) D_1, (g) Z_1, and (h) Z_3.

SUPPLEMENTAL PROBLEMS

1. Find the image of the triangle in Fig. 13-35 under

    (a) the translation that slides up 3 units.
    (b) the translation that slides to the right 2 units and up 1 unit.
    (c) reflection across the y-axis.
    (d) reflection across the x-axis.
    (e) rotation 90° counterclockwise around the origin.

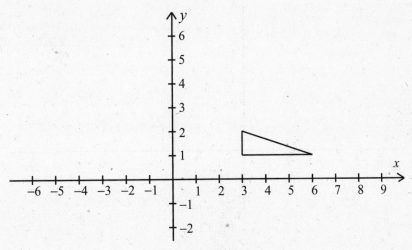

Fig. 13-35

2. Find the image of the semicircle in Fig. 13-36 under

    (a) translating to the right 9 units and up 3 units.
    (b) translating down by 3 units.
    (c) reflecting across the y-axis.
    (d) reflecting across the dotted line.
    (e) rotating clockwise 90° around the origin.
    (f) rotating clockwise 90° around the point (−2, 1).

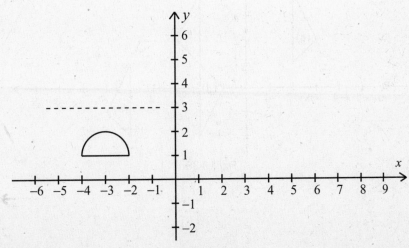

Fig. 13-36

3.　Find the image of the shape in Fig. 13-37 under

　(a)　translation by 3 units to the right and 1 unit down.
　(b)　reflection across the y-axis.
　(c)　reflection across the x-axis.
　(d)　reflection across the dotted line.
　(e)　rotation around the origin by 90° counterclockwise around the origin.
　(f)　rotation by 180° around the point (2, 2).

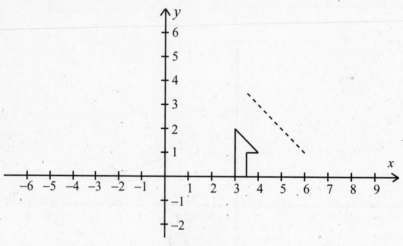

Fig. 13-37

4.　Find the transformation which takes triangle 1 to each of its images in Fig. 13-38.

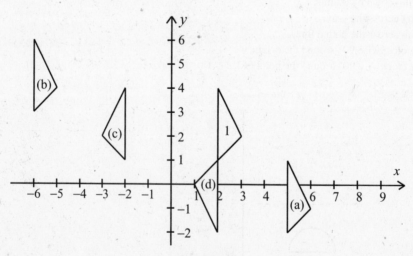

Fig. 13-38

5.   Find the transformation which takes the shape marked 1 in Fig. 13-39 to each of its images.

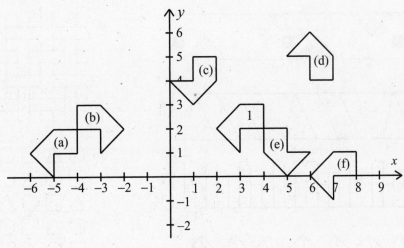

Fig. 13-39

6.   Name the sort of transformation (translation, reflection, rotation, or glide reflection), if any, which takes the arrow marked 1 to each of the other figures in Fig. 13-40.

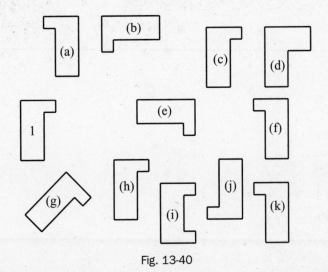

Fig. 13-40

7. Suppose each of the objects in Fig. 13-41 is infinite. Draw arrows to indicate the directions of translation symmetry, if any.

(a)

(b)

(c)

(d)

(e)

(f)

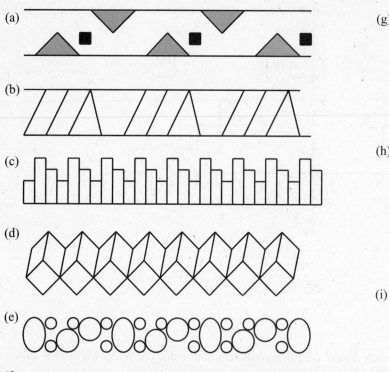

(g)

(h)

(i)

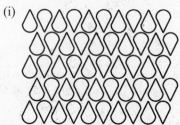

Fig. 13-41

8. Identify the symmetry group for each of the objects in Fig. 13-42. Draw all the axes for objects with reflection symmetry.

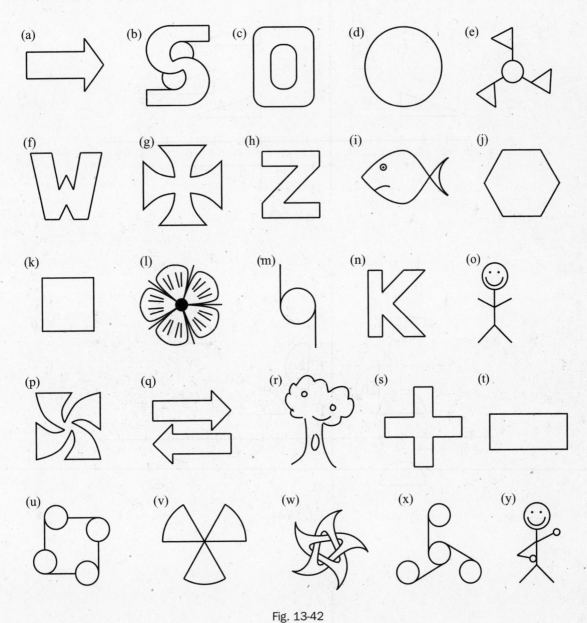

Fig. 13-42

Answers

1. The answers are shown in Fig. 13-43.
2. The answers are illustrated in Fig. 13-44.
3. The answers are illustrated in Fig. 13-45.
4. (a) Translation to the right 3 units and down 3 units
    (b) Translation to the left 8 units and up 2 units
    (c) Reflection across the y-axis
    (d) Rotation 180° around the point (2, 1)

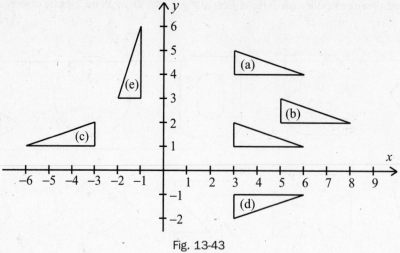

Fig. 13-43

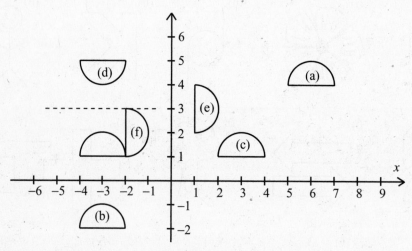

Fig. 13-44

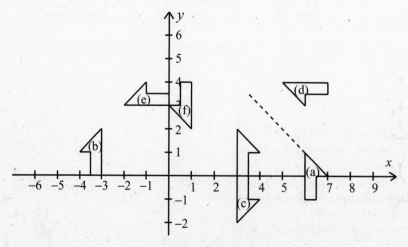

Fig. 13-45

5. (a) Translation to the left 8 units and down 1 unit
   (b) Reflection across the y-axis
   (c) Reflection across the diagonal line $y=x+1$
   (d) Rotation by 90° clockwise around the point (6, 2)
   (e) Rotation by 90° counterclockwise around the point (4, 2)
   (f) Translation to the right 4 units and down 2 units

6. (a) Glide reflection
   (b) Reflection
   (c) Translation
   (d) Not a transformed image
   (e) Rotation
   (f) Reflection
   (g) Rotation
   (h) Translation
   (i) Not a transformed image
   (j) Rotation
   (k) Glide reflection

7. The answers are given in Fig. 13-46.

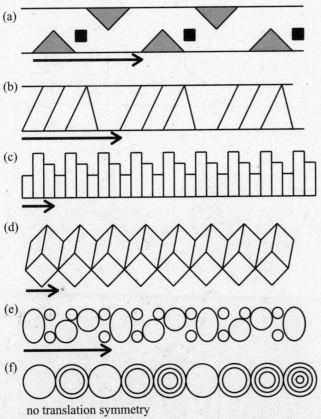

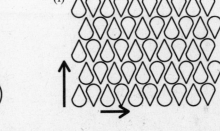

no translation symmetry

Fig. 13-46

8. The axes of reflection are shown in Fig. 13-47. (a) D_1, (b) Z_2, (c) D_2, (d) D_∞, (e) Z_3, (f) D_1, (g) D_4, (h) Z_2, (i) Z_1, (j) D_6, (k) D_4, (l) D_5, (m) Z_2, (n) D_1, (o) D_1, (p) Z_4, (q) Z_2, (r) Z_1, (s) D_4, (t) D_2, (u) Z_4, (v) D_3, (w) Z_5, (x) Z_3, and (y) Z_1.

Fig. 13-47

CHAPTER 14

Iterative Processes

When a simple action is repeated over and over again, it is called an *iterative process*. When a drawing is changed repeatedly using the same rule, the result is often a fractal. When a single number is added or multiplied over and over, the results will form a sequence of numbers. In this chapter will we look at a few of these situations.

Fractals

A *fractal* is an object that has parts which resemble the entire object. For example, it is conceivable that a tree might have a branch which looked like the whole tree itself, as illustrated in Fig. 14-1. Such a tree would be a fractal.

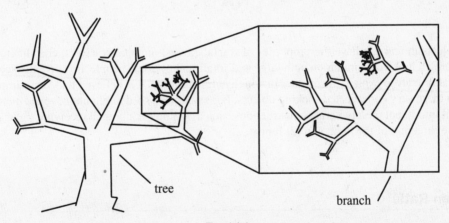

tree

branch

Fig. 14-1

Sierpinski's gasket is a fractal which contains quarter-size copies of itself at every level, as illustrated in Fig. 14-2. To draw this fractal, start with any triangle and connect the midpoints of its three sides, as shown in Fig. 14-3(a). Next, do the same to each of the smaller triangles except for the one in the middle, as shown in Fig. 14-3(b). The next iteration of this process is shown in Fig. 14-3(c).

Another example of a fractal is **R** = { all real numbers }. Every real number can be described by a number, perhaps with an infinite string of decimal digits. Now remove the first nonzero digit from each number. For example, the number 45.28 will become 5.28, the number 107.2 will become 7.2, and the number 0.02 will become 0. Only zero will have no nonzero digits, so we can leave that alone. When this is done to all the real numbers, the result will be many copies of all the real numbers. Any number, for example 42.7, will have come from many sources: 142.7, 342.7, 80042.7, etc. In some sense, we are looking inside the set of all real numbers and seeing many copies of itself. By definition, this means that the real numbers form a fractal.

Fig. 14-2

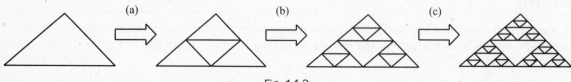

Fig. 14-3

Many objects in real life have the properties of fractals; trees, mountains, rivers, coastlines, blood vessels, and more all look very similar on both large and small scales. Some people are currently seeking to use mathematical fractals to explain the manner in which nature unfolds. It is believed that very simple rules are able to form these very complicated-looking objects. For example, a seed might not have the plans of the entire tree built into it, but merely iterative instructions such as (1) on each branch receiving light, grow a new branch in the direction of the light; and (2) repeat.

The Golden Ratio

Back in ancient times, mathematicians imagined that two lengths x and y would form an ideal ratio if the relationship between x and y was the same as that between y and $x+y$. This means that $\dfrac{x}{y} = \dfrac{y}{x+y}$. Such lengths would make the two triangles in Fig. 14-4 similar.

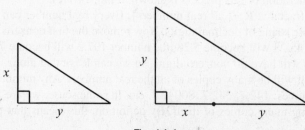

Fig. 14-4

Most numbers x and y do not satisfy $\dfrac{x}{y} = \dfrac{y}{x+y}$. For example, if $x=1$ and $y=3$, then $\dfrac{x}{y} = \dfrac{1}{3}$ and $\dfrac{y}{x+y} = \dfrac{3}{4}$. However, if $x=1$, then $\dfrac{x}{y} = \dfrac{y}{x+y}$ becomes $\dfrac{1}{y} = \dfrac{y}{1+y}$, which can be cross-multiplied to $1+y=y^2$ and set equal to zero: $y^2-y-1=0$. Using the quadratic formula with $a=1$, $b=-1$, and $c=-1$, we get $y = \dfrac{-b \pm \sqrt{b^2-4ac}}{2a} = \dfrac{-(-1) \pm \sqrt{(-1)^2-4(1)(-1)}}{2(1)} = \dfrac{1 \pm \sqrt{5}}{2}$. Because $\dfrac{1-\sqrt{5}}{2}$ is a negative number, it cannot represent the length y. Thus, $y = \dfrac{1+\sqrt{5}}{2} \approx 1.618$.

The ancient Greeks called the ratio, $\dfrac{y}{x} = \dfrac{1+\sqrt{5}}{2}$, the *golden ratio* and represented it by the letter phi: φ. They believed that the most beautifully proportioned rectangle, the *golden rectangle*, would have sides that formed this ratio. Because $\dfrac{x}{y} = \dfrac{y}{x+y}$, this means that both of the rectangles in Fig. 14-5(a) are golden rectangles.

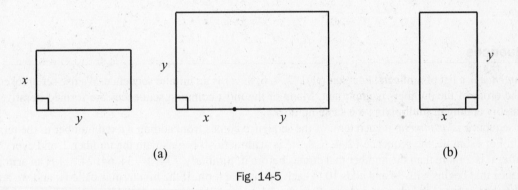

(a)   (b)

Fig. 14-5

Notice that if the small golden rectangle in Fig. 14-5 is rotated 90° (as shown in Fig. 14-5(b)), it will fit neatly into the larger golden rectangle. The other half of the rectangle will be a square, as illustrated in Fig. 14-6(a).

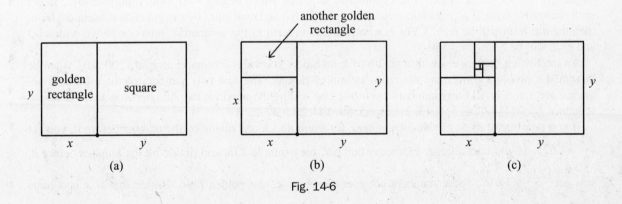

(a)   (b)   (c)

Fig. 14-6

Because the smaller rectangle is also golden, it can also be divided into a square and a golden rectangle, as illustrated in Fig. 14-6(b). This procedure can be repeated indefinitely, resulting in squares of ever-smaller sizes. If a quarter-circle is placed inside each of these squares in the right way, the result will be the *logarithmic spiral* illustrated in Fig. 14-7. Notice that the figures in Fig. 14-7 are fractals.

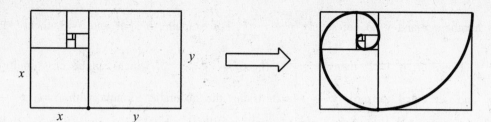

Fig. 14-7

Another way to write the golden ratio is with an infinite number of square roots inside of each other (called *nested radicals*): $\varphi = \sqrt{1+\sqrt{1+\sqrt{1+\sqrt{1+...}}}}$. In order to see this, set $y = \sqrt{1+\sqrt{1+\sqrt{1+\sqrt{1+...}}}}$, and then square both sides. This will result in $y^2 = 1+\sqrt{1+\sqrt{1+\sqrt{1+...}}}$, which is $y^2 = 1+y$, the same equation used earlier to obtain $y = \varphi$. Notice that this representation of φ is a fractal.

Sequences

A *sequence* is a list of numbers. For example, $\{2, 4, 6, 8,...\}$ is an infinite sequence. We use set brackets $\{\ \}$, but the order of the numbers is important. Many of the most common sequences are formed iteratively by repeatedly adding or multiplying the same number.

A sequence is *arithmetic* if each term of the sequence comes from adding a fixed number to the term before it. For example, the sequence $\{2, 4, 6, 8,...\}$ is arithmetic: it begins with the number 2, and every number after it is 2 more than the number that comes before it. Similarly, $\{34, 44, 54, 64, 74,...\}$ is an arithmetic sequence that begins with 34 and adds 10 to each successive term. If the first term is called a and we add the number b to it for each subsequent term, then the nth term will be $a+(n-1)b$. For example, the arithmetic sequence $\{13, 6, -1, -8, -15,...\}$ starts with $a=13$ and adds -7 for each subsequent term. Thus, the nth term will be $13+(n-1)(-7)$. For example, the fifth term will be $13+(5-1)(-7)=13-28=-15$, and the 100th term will be $13+(100-1)(-7)=-680$.

A sequence is *geometric* if each term is the result of multiplying the last number by a constant. For example, $\{5, 10, 20, 40, 80, 160,...\}$ is a geometric sequence which begins with 5 and multiplies by 2 to get each successive term. If a geometric sequence begins with a and multiplies by r to get each subsequent term, then the nth term will be $a \cdot r^{n-1}$. For example, the tenth term of the geometric sequence above with $a=5$ and $r=2$ will be $5 \cdot 2^{10-1}=2,560$.

An interesting sequence was thought up by a man now known as Fibonacci around 1200 A.D., when he imagined a problem concerning the reproduction of rabbits. The first two numbers of the Fibonacci sequence are 1 and 1. All other numbers are formed by adding the previous two numbers together. Thus, the sequence goes $\{1, 1, 2, 3, 5, 8, 13, 21, 34, 55, 89, 144, 233, 377,...\}$.

If you take one of these *Fibonacci numbers*, for example 13, and divide by the number before it, you get $\frac{13}{8}=1.625$. If you take a larger Fibonacci number, for example 233, and divide by the number before it, you get $\frac{233}{144} \approx 1.618$. These numbers get ever closer to φ, the golden ratio. Notice that if x is a large Fibonacci number and y is the number that comes after it, then the number after y will be $x+y$. If these ratios are close to the same number, then those numbers will satisfy $\frac{y}{x} \approx \frac{y+x}{y}$, which is equivalent to the equation which defines the golden ratio: $\frac{x}{y}=\frac{y}{x+y}$.

SOLVED PROBLEMS

Sequences

1. Name the first ten terms for each of the following sequences:
    (a) $\{2, 4, 6, 8, \ldots\}$
    (b) $\{5, 10, 20, 40, \ldots\}$
    (c) $\{1, 2, 3, 4, \ldots\}$
    (d) $\{5, 10, 15, 20, \ldots\}$
    (e) $\{1, -1, 1, -1, \ldots\}$
    (f) $\{1, 0, -1, 0, 1, 0, \ldots\}$
    (g) $\left\{1, \dfrac{1}{2}, \dfrac{2}{3}, \dfrac{3}{4}, \dfrac{4}{5}, \ldots\right\}$
    (h) $\{2, 3, 5, 7\}$

2. For each of the sequences in question (1), say if it is arithmetic, geometric, or neither.

3. Is it possible for a sequence to be geometric and arithmetic at the same time?

    ### Answers

    1. (a) 2, 4, 6, 8, 10, 12, 14, 16, 18, and 20
        (b) 5, 10, 20, 40, 80, 160, 320, 640, 1,280, and 2,560
        (c) 1, 2, 3, 4, 5, 6, 7, 8, 9, and 10
        (d) 5, 10, 15, 20, 25, 30, 35, 40, 45, and 50
        (e) 1, −1, 1, −1, 1, −1, 1, −1, 1, and −1
        (f) 1, 0, −1, 0, 1, 0, −1, 0, 1, and 0

        (g) $1, \dfrac{1}{2}, \dfrac{2}{3}, \dfrac{3}{4}, \dfrac{4}{5}, \dfrac{5}{6}, \dfrac{6}{7}, \dfrac{7}{8}$, and $\dfrac{8}{9}$

        (h) 2, 3, 5, and 7. This sequence has only four terms.

    2. (a) arithmetic with $a = 2$ and $b = 2$
        (b) geometric with $a = 5$ and $r = 2$
        (c) arithmetic with $a = 1$ and $b = 1$
        (d) arithmetic with $a = 5$ and $b = 5$
        (e) geometric with $a = 1$ and $b = -1$
        (f) This sequence could not be geometric, because nothing could come after 0 except 0 in a geometric sequence. This sequence could not be arithmetic, because the sum could never go from positive to negative and then back to positive again, not if every change were adding or subtracting the same constant each time.
        (g) This sequence is neither arithmetic nor geometric.

    3. Yes: any constant sequence, for example $\{5, 5, 5, 5, 5, \ldots\}$, is arithmetic, with $b = 0$, and geometric, with $r = 1$.

Series

Given a sequence $\{a_1, a_2, a_3, \ldots a_n, \ldots\}$, by the corresponding *series* we mean the sequence of partial sums $\{S_1, S_2, S_3, \ldots\}$ defined by $S_1 = a_1$, $S_2 = a_1 + a_2$, $S_3 = a_1 + a_2 + a_3, \ldots$, $S_n = a_1 + a_2 + a_3 + \ldots + a_n$, etc.

When the great German mathematician Carl Gauss was in grammar school, his teacher assigned everyone in the class to add up all the numbers from 1 to 100. The teacher had not left the room when Gauss declared his answer: 5,050. What he had done was to add the series to itself, but backward. By matching the 1 from the first series to the 100 of the second, the 2 with the 99, and so on as illustrated in Fig. 14-8, Gauss was able to add up one hundred 101's, for a total of $100 \times 101 = 100,100$. Because this was formed by

$$
\begin{array}{ccccccccc}
1 & + & 2 & + & 3 & + & \ldots & + & 100 \\
100 & + & 99 & + & 98 & + & \ldots & + & 1 \\
\hline
101 & + & 101 & + & 101 & + & \ldots & + & 101
\end{array}
$$

Fig. 14-8

adding two copies of the series, a single copy of the series was thus: $(1+2+3+\ldots+100)=\dfrac{10,100}{2}=5,050.$

In general, if an arithmetic sequence begins with a and adds b to form each subsequent term, then the sum of the first n terms of the sequence (an *arithmetic series*) will be $(a+(a+b)+(a+2b)+(a+3b)+\ldots(a+(n-1)b)=\dfrac{n(2a+(n-1)b)}{2}$. For example, the sequence $\{34,\ 44,\ 54,\ 64,\ 74,\ldots\}$ begins with $a=34$ and has $b=10$; so the sum of the first $n=5$ terms ought to be $\dfrac{5(2\cdot 34+(5-1)\cdot 10)}{2}=270$, as can be verified by actually adding: $34+44+54+64+74=270$.

If the first n terms of a geometric sequence are added (a *geometric series*), then the sum will be $S=a+ar+ar^2+ar^3+\ldots ar^{n-1}$. If this is multiplied by r, the result is $S\cdot r=ar+ar^2+ar^3+\ldots ar^n$. If we subtract these, most of the terms will cancel: $S-S\cdot r=a-ar^n$. This equation can be factored into $S(1-r)=a(1-r^n)$. This can then be solved $S=\dfrac{a(1-r^n)}{1-r}$.

As an example, the sum of the first six terms of the geometric sequence $\{5,\ 10,\ 20,\ 40,\ 80,\ 160,\ldots\}$ can be found using $S=\dfrac{a(1-r^n)}{1-r}$ with $a=5$, $r=2$, and $n=6$. The result is $5+10+20+40+80+160=\dfrac{5(1-2^6)}{1-2}=315$, as can be easily verified.

Incidentally, the financial mathematics formulas in Chapter 8 were obtained by using these formulas for arithmetic and geometric series.

SOLVED PROBLEMS

Series

1. Find the sum of the first 500 natural numbers: $1+2+3+4+\ldots+500$.
2. Find the sum of the arithmetic series: $3+8+13+18+23+28+\ldots 103$.
3. Find the sum of the first ten terms of this geometric sequence: $10, 30, 90, 270, \ldots$
4. Find the sum of the geometric series: $3+6+12+24+48+\ldots+12,288$.

Answers

1. This is an arithmetic series with $a=1$, $b=1$, and $n=500$. Thus, by the formula $\dfrac{n(2a+(n-1)b)}{2}$, the sum is $\dfrac{500(2(1)+(499)(1))}{2}=125,250$.

2. Here $a=3$, $b=5$, and the last term 103 must be $a+(n-1)b$; so $103=3+(n-1)\cdot 5$, thus 103 must be the $n=21$ term. Using the formula, the sum must be $\dfrac{21(2(3)+(20)\cdot 5)}{2}=1,113$.

3. Here $a=10$, $r=3$, and $n=10$; so using the formula $\dfrac{a(1-r^n)}{1-r}$, the sum must be $\dfrac{10(1-3^{10})}{1-3}=295,240$.

4. This is a geometric series with $a=3$ and $r=2$. To figure out how many terms we have, we need to solve $12{,}288 = a \cdot r^{n-1}$ for n. This leads to $12{,}288 = 3 \cdot 2^{n-1}$ and $4{,}096 = 2^{n-1}$. Because $2^{12} = 4{,}096$, this means $n=13$. The sum is thus $\dfrac{a(1-r^n)}{1-r} = \dfrac{3(1-2^{13})}{1-2} = 24{,}573$.

SUPPLEMENTAL PROBLEMS

1. Name the first ten terms of the following sequences. If the sequence is arithmetic, state a and b. If the sequence is geometric, state a and r.

   (a) $\{3, 6, 9, 12, 15, \ldots\}$
   (b) $\{13, 15, 17, 19, 21, \ldots\}$
   (c) $\{4, 8, 16, 32, 64, 128, \ldots\}$
   (d) $\{17, 12, 7, 2, -3, -8, -13, -18, \ldots\}$
   (e) $\{2, -6, 18, -54, 162, -486, \ldots\}$
   (f) $\{1, 2, 1, 3, 1, 4, \ldots\}$
   (g) $\{1, 1, 1, 1, 1, 1, 1, \ldots\}$

2. For each of the sequences for question (1), name the 15th term.
3. Find the sum of the first 80 natural numbers: $1 + 2 + 3 + 4 + \ldots + 80$.
4. Find the sum of the first 20 terms of this sequence: $5 + 10 + 15 + 20 + 25 + \ldots$
5. Find the sum of the first ten terms of this sequence: $8 + 4 + 2 + 1 + \dfrac{1}{2} + \dfrac{1}{4} + \ldots$
6. Find the sum: $7 + 15 + 23 + 31 + 39 + \ldots + 87$.
7. Find the sum: $2 + 20 + 200 + 2{,}000 + \ldots + 2{,}000{,}000$.
8. Find the sum: $5 + 10 + 20 + 40 + 80 + 160 + \ldots + 1{,}280$.

Answers

1. (a) 3, 6, 9, 12, 15, 18, 21, 24, 27, and 30; arithmetic with $a=3$ and $b=3$
   (b) 13, 15, 17, 19, 21, 23, 25, 27, 29, and 31; arithmetic with $a=13$ and $b=2$
   (c) 4, 8, 16, 32, 64, 128, 256, 512, 1,024, and 2,048; geometric with $a=4$ and $r=2$
   (d) 17, 12, 7, 2, -3, -8, -13, -18, -23, and -28; arithmetic with $a=17$ and $b=-5$
   (e) 2, -6, 18, -54, 162, -486, 1,458, -4,374, 13,122, and -39,366; geometric with $a=2$ and $r=-3$
   (f) 1, 2, 1, 3, 1, 4, 1, 5, 1, and 6; neither arithmetic nor geometric
   (g) 1, 1, 1, 1, 1, 1, 1, 1, 1, and 1; arithmetic with $a=1$ and $b=0$ as well as geometric with $a=1$ and $r=1$

2. (a) 45
   (b) 41
   (c) 65,536
   (d) -53
   (e) 9,565,938
   (f) 1
   (g) 1

3. 3,240
4. 1,050
5. 15.984375
6. 517
7. 2,222,222
8. 2,555

Trigonometry

Trigonometry studies the measurements of triangles. When the measures of some sides and angles are known, the measures of the other sides and angles can be calculated.

Similar Triangles

When two triangles have the same three angles, they are called *similar triangles*. Similar triangles might not be the same size, but they will have the same shape and proportions. This means that the lengths of one triangle can be found by multiplying the lengths of the other by a *scale factor*.

For example, the two triangles in Fig. 15-1 are similar because they have the same three angles: 20°, 60°, and 100°. The side between the 60° and 20° angles on the second triangle is three times longer than the corresponding side on the first triangle. This means that the scale factor is 3. The length x must thus be $x = 3 \times 35 = 105$ mm and $y = 3 \times 88 = 264$ mm.

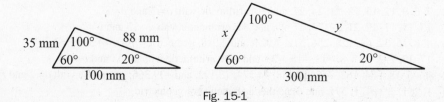

Fig. 15-1

The converse of this is also true: if the sides of one triangle are all scale multiples of another triangle, then the triangles are similar.

In Fig. 15-2, for example, the length of each side of the second triangle is half the length of the corresponding side in the first triangle. This means that the triangles are similar: $x = 70°$ and $y = 60°$. Because the angles of a triangle must sum to 180°, both z and w must measure 50°.

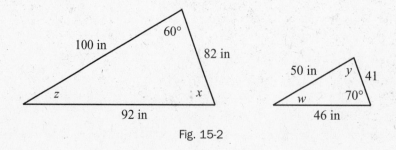

Fig. 15-2

SOLVED PROBLEMS

Similar Triangles

1. For each pair of triangles in Fig. 15-3, find the lengths x and y, and the angles a, b, and c, if possible.

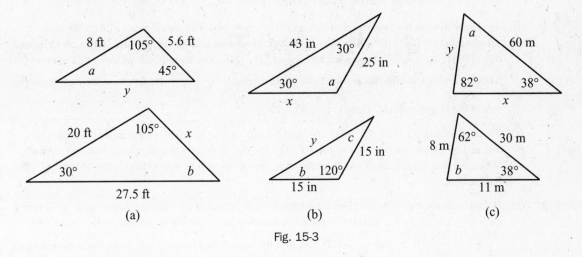

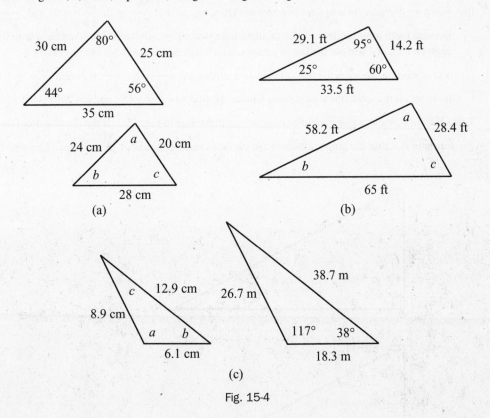

Fig. 15-3

2. Find the angles a, b, and c, if possible, using the triangles in Fig. 15-4.

Fig. 15-4

3. A 20-foot-tall streetlight casts a 15-foot shadow from a six-foot-tall man. How far is the man from the streetlight?

Answers

1. (a) Because the angles of a triangle must sum to 180°, we calculate that $a=180°-105°-45°=30°$ and $b=180°-105°-30°=45°$. Because the two triangles have the same three angles (30°, 45°, and 105°), they are similar. Looking at the two known corresponding sides, we see that each side of the larger triangle is $\frac{20}{8}=2.5$ times larger than the smaller triangle. Thus, $x=(2.5)\cdot5.6=14$ ft and $y=\frac{27.5}{2.5}=11$ ft.

   (b) The larger triangle has two angles of 30°, and thus is isosceles. This means that $x=25$ in. We also calculate that $a=180°-30°-30°=120°$. Similarly, the smaller triangle has two sides of length 15 in, and thus is isosceles: $b=c$. We thus calculate $b+b+120°=180°$, which can be solved: $b=30°$. The two triangles are thus similar. Each side of the smaller triangle is $\frac{15}{25}=\frac{3}{5}$, the length of the corresponding side of the bigger triangle. Thus, $y=\frac{3}{5}\cdot43=25.8$ in.

   (c) We calculate that $a=180°-82°-38°=60°$ and $b=180°-62°-38°=80°$. One triangle has angles 82°, 38°, and 60°, while the other has angles 80°, 38°, and 62°. It follows that the triangles are not similar. We cannot find the lengths of x and y at this time, though we will be able to later in this chapter with the law of sines.

2. (a) Before we know that these triangles are similar, we must check if the sides are proportional. By dividing each length from the big triangle by the corresponding length on the smaller triangle, we see that $\frac{30}{24}=1.25, \frac{25}{20}=1.25,$ and $\frac{35}{28}=1.25$. Because the scale factor for each side is the same, the triangles are similar. It follows that $a=80°$, $b=44°$, and $c=56°$.

   (b) When we compare the lengths of the two triangles, we see that $\frac{58.2}{29.1}=2$, $\frac{28.4}{14.2}=2$, and $\frac{65}{33.5}\approx1.94$. Because there is no single scale factor, these triangles are not similar. We will not be able to find the angles a, b, and c without the law of cosines, which will be covered later in this chapter.

   (c) The proportions between the sides of the two triangles are $\frac{38.7}{12.9}=3$, $\frac{18.3}{6.1}=3$, and $\frac{26.7}{8.9}=3$. Because these are all the same, the triangles are similar. Thus, $a=117°$, $b=38°$, and $c=180°-117°-38°=25°$.

3. This situation contains a pair of similar triangles, as illustrated in Fig. 15-5. If the distance from the man to the streetlight is x, then the proportions between the sides must be the same: $\frac{20}{6}=\frac{x+15}{15}$. Cross-multiplying

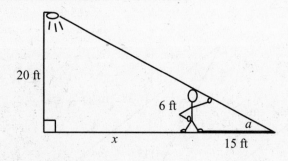

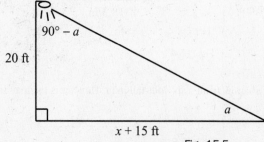

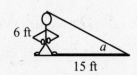

Fig. 15-5

results in $20 \cdot 15 = 6(x \times 15)$; thus, $300 = 6x + 90$. This means $x = 35$ ft. The man is thus 35 feet away from the base of the streetlight.

The Pythagorean Theorem

The longest side of a right triangle is called the *hypotenuse*, and the other two sides *legs*. If the legs have length a and b, and the hypotenuse has length c, as illustrated in Fig. 15-6, then the *Pythagorean theorem* states that $a^2 + b^2 = c^2$.

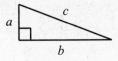

Fig. 15-6

Because a^2, b^2, and c^2 are the areas of squares with sides of length a, b, and c, this theorem is really a statement about areas, as illustrated in Fig. 15-7.

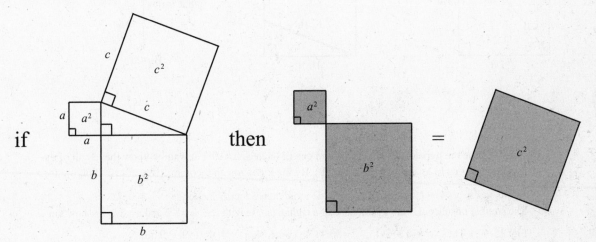

Fig. 15-7

It is easy to see that these areas are the same when four copies of the original triangle are added to each side of the equation, as shown in Fig. 15-8. This particular proof of the Pythagorean theorem comes from China and is at least 3,000 years old.

The converse of the Pythagorean theorem is also true: if the sides of a triangle are a, b, and c, where $a^2 + b^2 = c^2$, then the triangle will have a right angle.

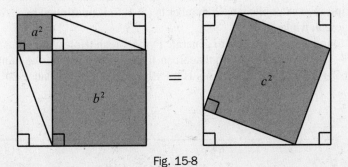

Fig. 15-8

SOLVED PROBLEMS

The Pythagorean Theorem

1. What is the diagonal measurement of a window that is three feet wide and five feet tall?
2. Suppose a 12-foot board is stuck into a shed that is only ten feet wide. If one end of the board rests on the floor, how high up the wall will the other end be?
3. To drive from the town of Blotin to North Millborough, one must drive 8.2 miles north and 3.7 miles east. What is the straight distance between the two towns?

Answers

1. The diagonal x of the window forms the hypotenuse c where the legs are $a=3$ and $b=5$, as shown in Fig. 15-9(a). By the Pythagorean theorem, $a^2+b^2=c^2$, which becomes $3^2+5^2=x^2$. Thus $x = \sqrt{34} \approx 5.83$ feet or approximately 5 feet, 10 inches.

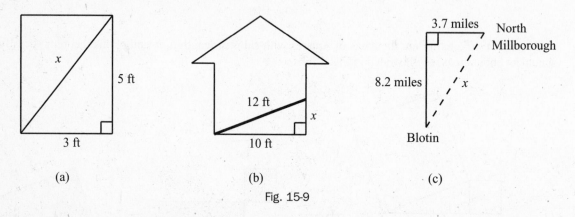

(a)   (b)   (c)

Fig. 15-9

2. Here we have the hypotenuse $c=12$ feet and one of the legs $a=10$, but want to know the length of the remaining leg $x=b$, as shown in Fig. 15-9(b). By the Pythagorean theorem, $10^2+x^2=12^2$. Solving for x results in $x = \sqrt{44} \approx 6.63$ feet. This is a little more than 6 feet, 7 inches.
3. The straight distance x is the hypotenuse of a right triangle with legs $a=8.2$ and $b=3.7$, as shown in Fig. 15-9(c). Thus, $x^2=(8.2)^2+(3.7)^2=80.93$. In conclusion, $x = \sqrt{80.93} \approx 9$ miles.

Pythagorean Triples

Almost as long as people have known the Pythagorean theorem, they have been interested in triples of natural numbers (a, b, c), where $a^2+b^2=c^2$. Such a triple of numbers is called a *Pythagorean triple*. The best-known Pythagorean triple is (3, 4, 5), but there are infinitely more.

If all three numbers of a Pythagorean triple are multiplied by the same natural number, the result will be another Pythagorean triple. For example, if each number in (3, 4, 5) is multiplied by 4, the result will be (12, 16, 20), another Pythagorean triple.

The ancient Greeks found another method to generate Pythagorean triples: if r and s are natural numbers with $r>s$, then $(2rs, r^2-s^2, r^2+s^2)$ will be a Pythagorean triple. For example, the numbers $r=5$ and $s=2$ generate $(2\cdot5\cdot2, 5^2-2^2, 5^2+2^2)= (20, 21, 29)$. This is a Pythagorean triple because $20^2+21^2= 400+441= 841=29^2$.

SOLVED PROBLEMS

Pythagorean Triples

Find the Pythagorean triples generated by the following:

1. $r = 3$ and $s = 2$
2. $r = 7$ and $s = 2$
3. $r = 6$ and $s = 1$
4. $r = 9$ and $s = 4$
5. $r = 27$ and $s = 20$

Answers

1. $(12, 5, 13)$
2. $(28, 45, 53)$
3. $(12, 35, 37)$
4. $(72, 65, 97)$
5. $(1{,}080, 329, 1{,}129)$

Trigonometric Functions

If x is a positive angle less than $90°$, it can be used as one angle of a right triangle, as illustrated in Fig. 15-10(a).

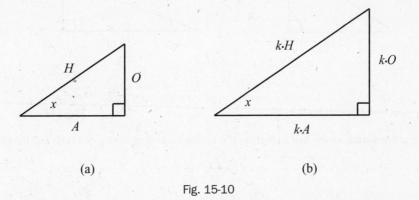

(a) (b)

Fig. 15-10

Traditionally, the hypotenuse is labeled H, the side adjacent to the angle x is labeled A, and the side opposite the angle x is labeled O. Any other right triangle with angle x will be similar to this one, and thus have sides of length $k \cdot H$, $k \cdot A$, and $k \cdot O$, where k is the scale factor, as illustrated in Fig. 15-10(b).

The *sine* of the angle x is the ratio $\dfrac{O}{H}$. Note that it does not matter which angle-x right triangle we use

because $\dfrac{k \cdot O}{k \cdot H}$ reduces to the same fraction. The *cosine* of the angle x is $\dfrac{A}{H}$, and the *tangent* of the angle

is $\dfrac{O}{A}$. When written as functions, these are as follows:

$$\sin(x) = \frac{O}{H}$$

$$\cos(x) = \frac{A}{H}$$

$$\tan(x) = \frac{O}{A}$$

Many people use the mnemonic SOH-CAH-TOA to remember these ratios.

The reciprocals of these ratios form the last three trigonometric functions: *cosecant, secant,* and *cotangent.*

$$\csc(x) = \frac{1}{\sin(x)} = \frac{H}{O}$$

$$\sec(x) = \frac{1}{\cos(x)} = \frac{H}{A}$$

$$\cot(x) = \frac{1}{\tan(x)} = \frac{A}{O}$$

SOLVED PROBLEMS

Trigonometric Functions

1. Use the triangles in Fig. 15-11 to find all six trigonometric functions for angle (a) *x*, (b) *y*, and (c) *z*.

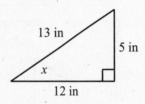 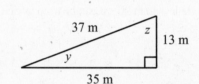

Fig. 15-11

Answers

1. (a)   The hypotenuse of the first triangle is $H=13$, the side adjacent to x is $A=12$, and the side opposite x is

$O=5$. Thus, $\sin(x) = \dfrac{O}{H} = \dfrac{5}{13}$, $\cos(x) = \dfrac{A}{H} = \dfrac{12}{13}$, $\tan(x) = \dfrac{O}{A} = \dfrac{5}{12}$, $\csc(x) = \dfrac{H}{O} = \dfrac{13}{5}$,

$\sec(x) = \dfrac{H}{A} = \dfrac{13}{12}$, and $\cot(x) = \dfrac{A}{O} = \dfrac{12}{5}$.

(b)   The hypotenuse of the triangle with angle y is $H=37$, the side adjacent to y is $A=35$, and the side

opposite is $O=13$. Thus, $\sin(y) = \dfrac{O}{H} = \dfrac{13}{37}$, $\cos(y) = \dfrac{A}{H} = \dfrac{35}{37}$, $\tan(y) = \dfrac{O}{A} = \dfrac{13}{35}$,

$\csc(y) = \dfrac{H}{O} = \dfrac{37}{13}$, $\sec(y) = \dfrac{H}{A} = \dfrac{37}{35}$, and $\cot(y) = \dfrac{A}{O} = \dfrac{35}{13}$.

(c)   The hypotenuse of the triangle with angle z is $H=37$. The side adjacent to z is $A=13$ (which happens

to be the side opposite angle y). The side opposite angle z is $O=35$. Thus, $\sin(z) = \dfrac{O}{H} = \dfrac{35}{37}$,

$\cos(z) = \dfrac{A}{H} = \dfrac{13}{37}$, $\tan(z) = \dfrac{O}{A} = \dfrac{35}{13}$, $\csc(z) = \dfrac{H}{O} = \dfrac{37}{35}$, $\sec(z) = \dfrac{H}{A} = \dfrac{37}{13}$, and

$\cot(z) = \dfrac{A}{O} = \dfrac{13}{35}$.

Trigonometry on a Calculator

A calculator that has buttons labeled SIN, COS, and TAN (usually near the top) can be used to find the trigonometric ratios of any angle. For example, $\sin(62°) \approx 0.8829$. If your calculator says $\sin(62) \approx -0.73918$, then it is reading angles not in degrees but in a different format called *radians*. Similarly, a calculator that says $\sin(62) = 0.82708$ is reading angles in a format called *gradians*. On an inexpensive calculator, the angle format can be changed by pressing the DRG button until *DEG* or *D* is displayed on the screen. With a graphing calculator, press the MODE button and then scroll down to select degrees.

If a circle of radius 1 is centered at the vertex of an angle, then the measure of the angle in radians is the length of the arc inside the angle. Because the circumference of a circle with radius 1 is 2π, this is the radian equivalent of 360°, as illustrated in Fig. 15-12(a). Similarly, a straight angle of 180° measures π radians, as shown in Fig. 15-12(b). A right angle measures 90° or $\dfrac{\pi}{2}$ radians, as shown in Fig. 15-12(c).

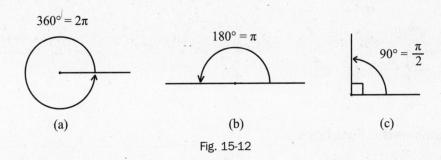

Fig. 15-12

In order to convert an angle from radians into degrees, multiply by $\dfrac{180°}{\pi}$. For example, an angle of $\dfrac{2\pi}{3}$ radians measures $\dfrac{2\pi}{3} \cdot \dfrac{180°}{\pi} = 120°$. Similarly, multiply by $\dfrac{\pi}{180°}$ to convert from degrees to radians. A 30° angle measures $30° \cdot \dfrac{\pi}{180°} = \dfrac{\pi}{6}$ radians.

SOLVED PROBLEMS

Trigonometry with a Calculator

1. Use a calculator to evaluate the following trigonometric functions. Round your answer to four decimal places.

    (a) $\sin(32°)$
    (b) $\cos(85°)$
    (c) $\tan(17.3°)$

2. Convert the following angles from radians into degrees:

    (a) $\dfrac{\pi}{4}$

    (b) $\dfrac{5\pi}{6}$

    (c) 1.27

3. Convert the following angles from degrees into radians:

    (a) 60°

    (b) 10°

    (c) 22.5°

Answers

1. (a) $\sin(32°) \approx 0.5299$
    (b) $\cos(85°) \approx 0.0872$
    (c) $\tan(17.3°) \approx 0.3115$

2. (a) $\dfrac{\pi}{4} = \dfrac{\pi}{4} \cdot \dfrac{180°}{\pi} = 45°$

    (b) $\dfrac{5\pi}{6} = \dfrac{5\pi}{6} \cdot \dfrac{180°}{\pi} = 150°$

    (c) $1.27 = (1.27) \cdot \dfrac{180°}{\pi} \approx \dfrac{(1.27) \cdot 180°}{3.14} \approx 72.8°$

3. (a) $60° = 60° \cdot \dfrac{\pi}{180°} = \dfrac{\pi}{3}$

    (b) $10° = 10° \cdot \dfrac{\pi}{180°} = \dfrac{\pi}{18}$

    (c) $22.5° = 22.5° \cdot \dfrac{\pi}{180°} = \dfrac{\pi}{8}$

Using Trigonometric Functions

If the measures of the angles and one side of a right triangle are known, then the lengths of the other sides can be found with trigonometry.

For example, the right triangle in Fig. 15-13 has a 32° angle and a width of eight feet. The eight-foot side is adjacent to the 32° angle, and the unknown side M is opposite. A trigonometric function which relates the opposite and adjacent sides is $\tan(x) = \dfrac{O}{A}$. Thus, $\tan(32°) = \dfrac{M}{8}$. Solving for M, we obtain $M = 8 \cdot \tan(32°) \approx 5$ ft.

To find the length of the hypotenuse N, we could use $\cos(x) = \dfrac{A}{H}$; thus, $\cos(32°) = \dfrac{8}{N}$. Solving for N results in $N = \dfrac{8}{\cos(32°)} \approx 9.4$ ft.

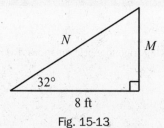

Fig. 15-13

The key to solving word problems with trigonometry is to imagine and draw the right triangle at the heart of the problem. Often the ground forms the base of the triangle and the wall, pole, tree, or similar object is assumed to rise up at a right angle. Once the right triangle has been drawn and all the relevant information has been labeled, solving the problem is only as difficult as the last example.

For example, suppose a ramp runs ten feet along the diagonal at a 14° angle. How high up does the ramp reach?

The triangle for this problem is illustrated in Fig. 15-14. As usual, the ground forms the base of the triangle and the height h rises up at a right angle. The two labeled sides are the side $O = h$ opposite the 14° angle and the hypotenuse $H = 10$. These sides are related by the sine function $\sin(x) = \dfrac{O}{H}$; thus, $\sin(14°) = \dfrac{h}{10}$. Solving for h, we obtain $h = 10 \cdot \sin(14°) \approx 2.4$ feet.

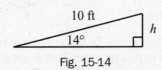

Fig. 15-14

SOLVED PROBLEMS

Using Trigonometric Functions

1. A plane takes off from the ground at a 12° angle. If it flies straight for 30,000 feet, how high off the ground will it be?
2. A ramp will be built at an 8° angle. If it is to reach a height of 3 feet, how long must it be along the diagonal? How long must it be in width?
3. A wire runs straight from the top of a pole to a point on the ground six feet from the base of the pole. If the wire makes a 78° angle with the ground, how tall is the pole?
4. A security light is built to shine up from the ground at a 60° angle. How high must it be attached to the outside of a house to shine at a spot 25 feet from the house?

Answers

1. The triangle for this problem is illustrated in Fig. 15-15(a). We know the hypotenuse is $H = 30,000$ and want to know the side h opposite the 12° angle, so we use $\sin(12°) = \dfrac{h}{30,000}$. When this is solved for h, the result is $h = 30,000 \cdot \sin(12°) \approx 6,237$ feet.

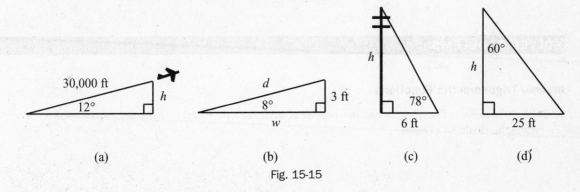

| (a) | (b) | (c) | (d) |

Fig. 15-15

2. Here we know the length opposite the 8° is three feet, and we want to know the length w of the adjacent side, as illustrated in Fig. 15-15(b). The tangent function is the one that relates the opposite and adjacent sides; thus, $\tan(8°) = \dfrac{3}{w}$. When this is solved for w, the result is $w = \dfrac{3}{\tan(8°)} \approx 21.3$ feet. To find the diagonal, we can use the sine function: $\sin(8°) = \dfrac{3}{d}$. When this is solved for d, the result is $d = \dfrac{3}{\sin(8°)} \approx 21.5$ feet.

3. As shown in Fig. 15-15(c), the side adjacent to the 78° angle is six feet long. To find the length of the opposite side h, we use $\tan(78°) = \dfrac{h}{6}$. Thus, $h = 6 \cdot \tan(78°) \approx 28.2$ feet.

4. In this problem, the side opposite the 60° angle is 25 feet long and the unknown length h is the adjacent side, as shown in Fig. 15-15(d). We use $\tan(60°) = \dfrac{25}{h}$ to find $h = \dfrac{25}{\tan(60°)} \approx 14.4$ feet.

Inverse Trigonometric Functions

A calculator can quickly compute the sine, cosine, or tangent of any angle. It can also undo this process and identify the angle that corresponds to a particular ratio. This is done with *inverse trigonometric functions*. For example, the function that undoes sine is called the *inverse sine*, written $\sin^{-1}$. If $\sin(x) = \dfrac{O}{H}$, then $\sin^{-1}\left(\dfrac{O}{H}\right) = x$. Similarly, if $\cos(x) = \dfrac{A}{H}$, then the *inverse cosine* $\cos^{-1}\left(\dfrac{A}{H}\right) = x$, and if $\tan(x) = \dfrac{O}{A}$, then the *inverse tangent* $\tan^{-1}\left(\dfrac{O}{A}\right) = x$. To invert a trigonometric function on a calculator, press either INV or 2ND before the trigonometric function button.

For example, the angle x illustrated in Fig. 15-16 has $\sin(x) = \dfrac{9}{16}$; thus, $\sin^{-1}\left(\dfrac{9}{16}\right) = x$. To evaluate this on a scientific calculator, press $9 \div 16 = \text{INV SIN} =$. On a graphing calculator, press INV SIN $9 \div 16 \,) =$. In either case, we will find the measure of the angle is $x \approx 34.2°$.

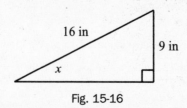

Fig. 15-16

SOLVED PROBLEMS

Inverse Trigonometric Functions

1. Use the triangles in Fig. 15-17 to find the measure of angle (a) x, (b) y, and (c) z. Round your answers to the nearest hundredth of a degree.

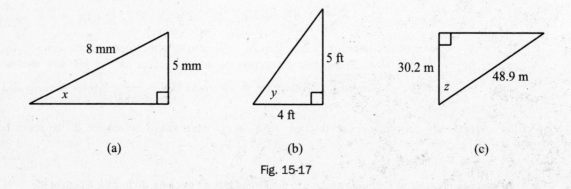

(a)　　　　　　　　(b)　　　　　　　　(c)

Fig. 15-17

Answers

1. (a) $\sin(x) = \dfrac{5}{8}$, so $x = \sin^{-1}\left(\dfrac{5}{8}\right) \approx 38.68°$

   (b) $\tan(y) = \dfrac{5}{4}$, so $y = \tan^{-1}\left(\dfrac{5}{4}\right) \approx 51.34°$

   (c) $\cos(z) = \dfrac{30.2}{48.9}$, so $z = \cos^{-1}\left(\dfrac{30.2}{48.9}\right) \approx 51.86°$

Using Inverse Trigonometric Functions

Most word problems that request the measure of an angle can be solved by using inverse trigonometric functions. As with all trigonometry problems, the trick is to draw and label the right triangle that is described in the problem.

For example, suppose a group of scouts has traveled 3,000 meters north and then 1,500 meters to the west. In what direction should they head to return straight to where they started?

To answer this problem, we draw the triangle in Fig. 15-18, label the lengths of the two known sides, and mark the unknown angle x. The tangent of this angle is $\tan(x) = \dfrac{3,000}{1,500} = 2$. Solving for x results in $x = \tan^{-1}(2) \approx 63.43°$. The scouts should thus walk about 63° south of east to return to where they started.

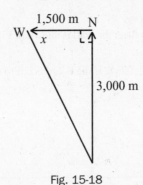

Fig. 15-18

SOLVED PROBLEMS

Using Inverse Trigonometric Functions

1. Suppose a ship's destination is 40 miles east and 14 miles north of its current location. In what direction should it head to go straight to its destination?
2. If 27 meters of wire run straight from the top of a 25-meter tower to the ground, what angle will it make with the ground?
3. The lights in a theatre are 20 feet off the ground. At what angle should a light be turned to aim at a spot on the ground that is seven feet from directly underneath it?
4. An eight-foot board will be attached to a spot six feet up on a wall in order to support it. At what angle should the top of the board be cut in order to lie flush with the wall?

Answers

1. The ship wants to travel along the diagonal of the triangle illustrated in Fig. 15-19(a). The angle x here has $\tan(x) = \dfrac{14}{40}$. This means that $x = \tan^{-1}\left(\dfrac{14}{40}\right) \approx 19.29°$. Thus, the ship should head 19.29° north of east to go straight to its destination.

2. The triangle for this problem is illustrated in Fig. 15-19(b). Because $\sin(x) = \dfrac{25}{27}$, we can calculate $x = \sin^{-1}\left(\dfrac{25}{27}\right) \approx 67.8°$.

3. The triangle for this problem is illustrated in Fig. 15-19(c). Here, $\tan(x) = \dfrac{7}{20}$. This is equivalent to $x = \tan^{-1}\left(\dfrac{7}{20}\right) \approx 19.29°$. The light should thus be aimed around 19° up from pointing straight at the ground.

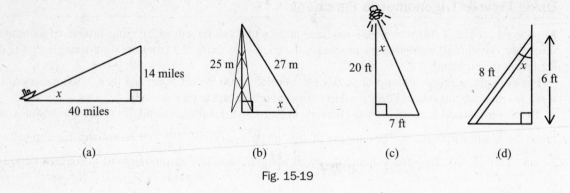

Fig. 15-19

4. The angle x at which the board will meet the wall is shown in Fig. 15-19(d). Here, $\cos(x) = \dfrac{6}{8}$. This is equivalent to $x = \cos^{-1}\left(\dfrac{6}{8}\right) \approx 41.4°$. Thus, if the board is cut to make a 41° angle at the tip, it will lie neatly against the wall.

The Law of Sines

The *law of sines* enables us to use trigonometry to calculate the angles and sides of nonright triangles. If the triangle has angles of measure A, B, and C opposite sides of length a, b, and c respectively, as illustrated in Fig. 15-20, then the law of sines states that $\dfrac{a}{\sin(A)} = \dfrac{b}{\sin(B)} = \dfrac{c}{\sin(C)}$.

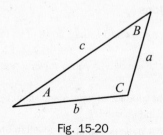

Fig. 15-20

For example, suppose we wanted to know the lengths x and y of the triangle illustrated in Fig. 15-21. Because the angle opposite x is 38°, we can say that $a = x$ and $A = 38°$. Similarly, we can label $b = 60$ and $B = 82°$. By the law of sines, $\dfrac{a}{\sin(A)} = \dfrac{b}{\sin(B)}$, and thus $\dfrac{x}{\sin(38°)} = \dfrac{60}{\sin(82°)}$. It follows that $x = \dfrac{60\sin(38°)}{\sin(82°)} \approx 37.3$ meters.

If we want to find $y = c$, we will need to calculate the measure of the opposite angle $C = 180° - 82° - 38° = 60°$. Because $\dfrac{b}{\sin(B)} = \dfrac{c}{\sin(C)}$ by the law of sines, we know that $\dfrac{60}{\sin(82°)} = \dfrac{y}{\sin(60°)}$. It follows that $y = \dfrac{60\sin(60°)}{\sin(82°)} \approx 52.5$ meters.

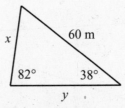

Fig. 15-21

SOLVED PROBLEMS

The Law of Sines

1. Find the lengths of sides (a) x, (b) y, and (c) z of the triangles in Fig. 15-22.

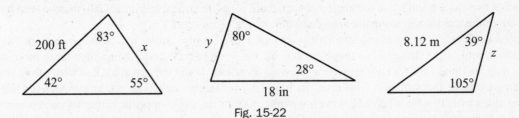

Fig. 15-22

2. You run a kingdom with two towers. The distance between your two towers is 500 yards. The first tower to catch sight of your enemy marks his army as 87° from the second tower. Immediately afterward, the second tower marks the flag of the general as 72° from the first tower. How far is the enemy from each of your towers?

Answers

1. (a) In the first triangle, we can set $a=x$, $A=42°$, $b=200$, and $B=55°$, and use the law of sines

$$\frac{a}{\sin(A)} = \frac{b}{\sin(B)}.$$ This makes $\frac{x}{\sin(42°)} = \frac{200}{\sin(55°)}$. Thus, $x = \frac{200\sin(42°)}{\sin(55°)} \approx 163.37$ ft.

   (b) In the second triangle, we can set $a=y$, $A=28°$, $b=18$, and $B=80°$, and use the law of sines

$$\frac{a}{\sin(A)} = \frac{b}{\sin(B)}.$$ This means $\frac{y}{\sin(28°)} = \frac{18}{\sin(80°)}$; thus, $y = \frac{18\sin(28°)}{\sin(80°)} \approx 8.58$ in.

   (c) To use the law of sines on the third triangle, we will need to calculate the angle opposite the unknown z. If $a=z$, then $A=180°-39°-105°=36°$. We can then set $b=8.12$ and $B=105°$. Thus,

$$\frac{z}{\sin(36°)} = \frac{8.12}{\sin(105°)} \text{ and } z = \frac{8.12\sin(36°)}{\sin(105°)} \approx 4.94 \text{ m}.$$

2. The illustration for this problem is in Fig. 15-23. The first thing to see is that we will need the angle opposite the 500-yard length: $180°-87°-72°=21°$. The law of sines states that $\frac{500}{\sin(21°)} = \frac{x}{\sin(72°)}$.

   Thus, $x = \frac{500\sin(72°)}{\sin(21°)} \approx 1,327$ yards away from the first tower. Similarly, the second tower is

$$y = \frac{500\sin(87°)}{\sin(21°)} \approx 1,393 \text{ yards away}.$$

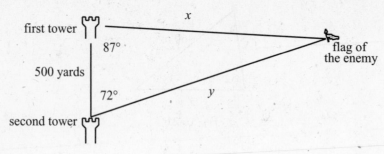

Fig. 15-23

The Law of Cosines

The law of sines only can be used if we know the measures of a side and the angle opposite it. If we know the lengths of two sides a and b and the angle between them C (as illustrated in Fig. 15-20), then we can find the measure of the third side c using the *law of cosines*: $a^2 + b^2 - 2ab \cdot \cos(C) = c^2$.

For example, suppose we have a group of scouts who travel for two miles in a certain direction. They see something cool in the distance, so they turn 60° off the straight path and continue for three more miles. When they are done, how far have they gone? In what direction should they head back to where they started?

Here, we have the situation illustrated in Fig. 15-24. We can calculate the angle between the two known sides as $180° - 60° = 120°$. However, we do not know an angle opposite either of our known sides so we cannot use the law of sines. Fortunately, the law of cosines will find the length of the opposite side c using $a = 2$, $b = 3$, and $C = 120°$. This results in $2^2 + 3^2 - 2(2)(3)\cos(120°) = c^2$. Solving for c results in $c = \sqrt{13 - 12\cos(120°)} \approx 4.36$ miles. We can now find the angle A using the law of sines: $\dfrac{2}{\sin(A)} = \dfrac{4.36}{\sin(120°)}$. From this, $\sin(A) = \dfrac{2\sin(120°)}{4.36}$ and thus $A = \sin^{-1}\left(\dfrac{2\sin(120°)}{4.36°}\right) \approx 23.4°$. The scouts should turn around to the left and head 23.4° from doubling exactly back in the direction from which they most recently came.

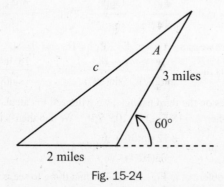

Fig. 15-24

The law of cosines can also be used to find angles for a nonright triangle when all three sides are known. For example, suppose you have three pieces of plywood (one four ft long, one six feet long, and one eight feet long) with which you wanted to make a triangular shelter. Suppose further that knowing the exact angles of the three corners would help in devising the fasteners.

The overhead view of the imagined shelter is illustrated in Fig. 15-25. Using $a = 4$, $b = 6$, and $c = 8$ with the law of cosines results in $4^2 + 6^2 - 2(4)(6)\cos(A) = 8^2$, which can be solved to $\cos(A) = \dfrac{8^2 - 4^2 - 6^2}{-2(4)(6)}$.

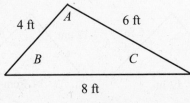

Fig. 15-25

This means $A = \cos^{-1}\left(\dfrac{64-14-36}{-48}\right) \approx 107°$. Now the law of sines can be used to find the angle B via

$\dfrac{6}{\sin(B)} = \dfrac{8}{\sin(107°)}$ to get $\sin(B) = \dfrac{6\sin(107°)}{8}$; thus, $B = \sin^{-1}\left(\dfrac{3\sin(107°)}{4}\right) \approx 46°$. The final angle

$C \approx 180° - 107° - 46° = 27°$.

SOLVED PROBLEMS

Law of Cosines

1. Find the sides x and y and the angle z using the triangles in Fig. 15-26.

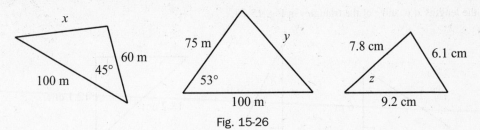

Fig. 15-26

Answers

1. To find x, we use the law of cosines $(a^2 + b^2 - 2ab\cos(C) = c^2)$, with $a = 60$, $b = 100$, $C = 45°$, and $c = x$. This results in $60^2 + 100^2 - 2(60)(100)\cos(45°) = x^2$. In conclusion,

$x = \sqrt{136,000 - 120,00\cos(45°)} \approx 71.52$ m.

To find y, we use the law of cosines $(75^2 + 100^2 - 2(75)(100)\cos(53°) = y^2)$. This means

$y = \sqrt{75^2 + 100^2 - 150(100)\cos(53°)} \approx 81.23$ m.

To find z, we use the law of cosines with $a = 7.8$, $b = 9.2$, $c = 6.1$, and $C = z$. This results in $7.8^2 + 9.2^2 - 2(7.8)(9.2)\cos(z) = 6.1^2$, which, when solved for z, results in

$z = \cos^{-1}\left(\dfrac{6.1^2 - 7.8^2 - 9.2^2}{-2(7.8)(9.2)}\right) \approx 41°$.

SUPPLEMENTAL PROBLEMS

1. The triangles in Fig. 15-27 are not drawn to scale. Use the given angles and lengths to determine which triangles are similar.
2. Use the triangles in Fig. 15-28 to find the lengths of the sides w, x, y, and z.

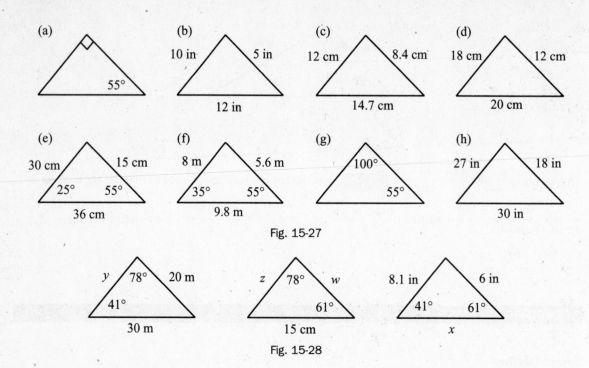

Fig. 15-27

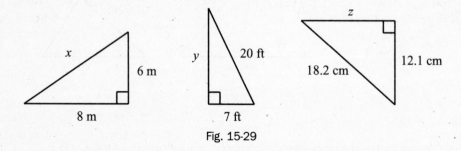

Fig. 15-28

3. A four-foot-tall child stands 18 feet away from a lamp hung up in a tree. If the girl's shadow is 12 feet long, how high up is the lamp?

4. Find the lengths x, y, and z of the triangles in Fig. 15-29.

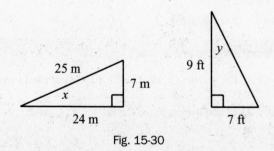

Fig. 15-29

5. A 30-foot ladder is leaned against a wall. If the bottom of the ladder is placed eight feet from the base of the wall, how high up the wall will the ladder touch?

6. A painting is three feet wide and 4.5 feet tall. What is its diagonal measure?

7. Which of the following triples of numbers are Pythagorean triples: (a) (8, 15, 17), (b) (5, 12, 15), (c) (9, 40, 41), (d) (39, 80, 89), and (e) (93, 165, 193)?

8. Using the ancient Greek method, find the Pythagorean triples generated by (a) $r=6$ and $s=5$, (b) $r=10$ and $s=7$, and (c) $r=15$ and $s=11$.

9. Use the triangles in Fig. 15-30 to find all six trigonometric functions of angles x and y.

Fig. 15-30

10. Use a calculator to evaluate the following trigonometric functions to four decimal places: (a) sin(71°), (b) cos(11°), (c) tan(42°), (d) sin(65.3°), (e) cos(19.2°), and (f) tan(80°).

11. Convert the following angles from radians into degrees: (a) $\dfrac{\pi}{5}$, (b) $\dfrac{3\pi}{4}$, (c) $\dfrac{7\pi}{6}$, (d) 0.523, and (e) 2.18/

12. Convert the following angles from degrees into radians: (a) 20°, (b) 270°, (c) 50°, (d) 67°, and (e) 16.4°.

13. A light is bolted to a stage at a distance of 24 feet from the backdrop. If the light aims upward at a 30° angle, how high up on the backdrop will the light be centered?

14. A steep stretch of road runs straight at a 10° inclination for five miles. What is the difference in elevation (measured in feet) between the two ends of this stretch of road?

15. A person flying in a plane is 20,000 feet above the ground. If she sees an object 25° below the horizontal, how far away is that object (measured along the diagonal)?

16. Find the measures of the angles x, y, and z illustrated in Fig. 15-31.

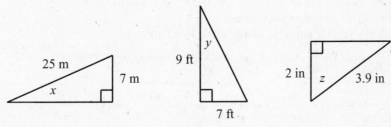

Fig. 15-31

17. A helicopter is summoned to an accident that has occurred 18 miles north and 3.4 miles west of its current location. In what direction should it head?

18. A 20-foot ladder leans against a wall. If the base of the ladder is three feet from the base of the wall, what angle does the ladder make with the ground?

19. A landscaper wants a plot of land to descend one inch for every ten feet of horizontal distance. What angle is this?

20. Find the lengths x, y, and z for the nonright triangles in Fig. 15-32.

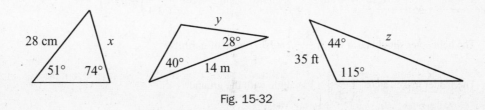

Fig. 15-32

21. Two fire towers are 40 miles apart. The first sights a fire 35° to the left of the second tower. The second tower sights the same fire at 52° from the first. How far is the fire from the first tower?

22. Find the lengths x and y and the angle z shown in Fig. 15-33.

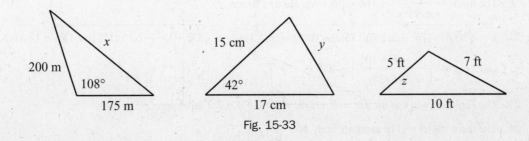

Fig. 15-33

23. A pair of explorers split up. One travels straight for eight miles. The other goes straight for seven miles in a direction 15° east of the first explorer. How far apart will they end up?

24. Three bars of steel are welded together into a triangle. If the lengths of the bars are 8, 11, and 12 feet, what will the three angles of the triangle be?

Answers

1. Triangles (a), (c), and (f) are all similar. Triangles (b), (e), and (g) are all similar. Triangles (d) and (h) are similar.

2. $w = 10\,\text{cm}$, $x = 9\,\text{in}$, $y = 27\,\text{m}$, and $z = 13.5\,\text{cm}$

3. The lamp is ten feet off the ground.

4. $x = 10\,\text{m}$, $y = \sqrt{351} \approx 18.7\,\text{ft}$, and $z = \sqrt{184.83} \approx 13.6\,\text{cm}$

5. The ladder will lean against a spot $\sqrt{836} \approx 28.9\,\text{ft} \approx 28$ feet, 11 inches up the wall.

6. The diagonal measure will be $\sqrt{29.25} \approx 5.4$ feet ≈ 5 feet, 5 inches.

7. (a), (c), and (d) are all Pythagorean triples.

8. (a) (11, 60, 61), (b) (51, 140, 149), and (c) (104, 330, 346)

9. $\sin(x) = \dfrac{7}{25}$, $\cos(x) = \dfrac{24}{25}$, $\tan(x) = \dfrac{7}{24}$, $\csc(x) = \dfrac{25}{7}$, $\sec(x) = \dfrac{25}{24}$, $\cot(x) = \dfrac{24}{7}$,

   $\sin(y) = \dfrac{7}{\sqrt{130}}$, $\cos(y) = \dfrac{9}{\sqrt{130}}$, $\tan(y) = \dfrac{7}{9}$, $\csc(y) = \dfrac{\sqrt{130}}{7}$, $\sec(y) = \dfrac{\sqrt{130}}{9}$, and $\cot(y) = \dfrac{9}{7}$

10. (a) 0.9455, (b) 0.981.6, (c) 0.9004, (d) 0.9085, (e) 0.9444, and (f) 5.6713

11. (a) 36°, (b) 135°, (c) 210°, (d) $\dfrac{(0.523) \cdot 180°}{\pi} \approx 30°$, and (e) $\dfrac{(2.18) \cdot 180°}{\pi} \approx 124.9°$

12. (a) $\dfrac{\pi}{9}$, (b) $\dfrac{3\pi}{2}$, (c) $\dfrac{5\pi}{18}$, (d) $\dfrac{67\pi}{180}$, and (e) $\dfrac{(16.4)\pi}{180}$

13. The light is aimed $24\tan(30°) \approx 13.8$ feet ≈ 13 feet, 10 inches up the backdrop.

14. The road ascends $5\sin(10°) \approx 0.87$ miles $\approx 4{,}584$ feet.

15. The object is $\dfrac{20{,}000}{\sin(25°)} \approx 47{,}324$ feet away, measured along the diagonal.

16. $x = \sin^{-1}\left(\dfrac{7}{25}\right) \approx 16.26°$, $y = \tan^{-1}\left(\dfrac{7}{9}\right) \approx 37.87°$, and $z = \cos^{-1}\left(\dfrac{2}{3.9}\right) \approx 59.15°$

17. The helicopter should head $\tan^{-1}\left(\dfrac{3.4}{18}\right) \approx 10.7°$ west of north.

18. The ladder makes a $\cos^{-1}\left(\dfrac{3}{20}\right) \approx 81°$ angle with the ground.

19. The ground will descend $\tan^{-1}\left(\dfrac{1}{120}\right) \approx 0.5°$.

20. $x = \dfrac{28\sin(51°)}{\sin(74°)} \approx 22.6$ cm, $y = \dfrac{14\sin(40°)}{\sin(112°)} \approx 9.7$ m, and $z = \dfrac{35\sin(115°)}{\sin(21°)} \approx 88.5$ ft

21. The fire is $\dfrac{40\sin(52°)}{\sin(93°)} \approx 31.6$ miles from the first tower.

22. $x = \sqrt{200^2 + 175^2 - 2(200)(175)\cos(108°)} \approx 303.7$ m, $y = \sqrt{15^2 + 17^2 - 2(15)(17)\cos(42°)} \approx 11.6$ cm, and $z = \cos^{-1}\left(\dfrac{7^2 - 5^2 - 10^2}{-2(5)(10)}\right) \approx 40.5°$

23. The explorers will end up $\sqrt{8^2 + 7^2 - 2(8)(7)\cos(15°)} \approx 2.2$ miles apart.

24. The three angles will be approximately 40°, 63°, and 77°.

Index